DESERT
AIRLINERS

DESERT
AIRLINERS

GRAHAM ROBSON

Motorbooks International
Publishers & Wholesalers ®

ACKNOWLEDGEMENTS

I am indebted to Sgt Dawn J. McKee at Davis-Monthan AFB Public Relations Office for all her help in arranging access to the AMARC, together with Dan Sabovitch and his staff at Mojave Airport, and Veronica Jiminez at Evergreen Air Center, Marana, for all their assistance. With a special thank you to a good friend, Mr A. E. "Schnozz" Mayer, at Marana, for all his time, co-operation and generosity over the years. All of whom made this book possible.

This edition first published in 1994 by Motorbooks International, Publishers & Wholesalers, PO Box 2, 729 Prospect Avenue, Osceola, WI 54020, USA.

© Graham Robson, 1994
Reprinted 1996

Previously published by Airlife Publishing Ltd, Shrewsbury, England, 1994

Library of Congress Cataloging-in-Publication Data is available
Robson, Graham
   Desert airliners / Graham Robson
     p.   cm.
   Includes index.
   ISBN 0-87938-904-4
   1. Airplanes – California – Mojave Desert – Pictorial works.
   2. Airplanes – California – Mojave Desert – Storage.
   I. Title.
   TL549.R633   1994
   629.133'340423 – dc20   93-49485
                                  CIP

Printed and bound in Singapore.

CONTENTS

INTRODUCTION

The idea of storing unwanted aircraft in the desert is nothing new, during the immediate post-war years huge surplus aircraft parks were created at numerous different sites throughout the central and western United States, as vast numbers of war weary bombers and fighters returned home.

As combat units returned to the United States, so the storage sites began to fill with their aircraft, to face an all too certain end. Amongst the thousands of combat veterans in storage were many aircraft which had been built too late to see active service and were simply flown directly into storage from their respective production 'finishing' plants pending disposal. Huge surplus sales soon followed, and the once full storage sites were gradually emptied by industrious scrap metal merchants, with hungry furnaces.

The sad spectacle of rows and rows of redundant aircraft has now returned to haunt a number of these former wartime surplus parks, only this time the inmates wear the bright colours of many well-known airlines. During the last decade or so, a number of locations in Arizona, Nevada and southern California have become well-known as temporary home for hundreds of retired airliners, the climatic conditions which prevail in this area making it most suitable for the long-term preservation of stored aircraft. The acres of parked aircraft also include a significant number of brand new 'factory fresh' airframes which were flown directly into desert storage from the production line. This bizarre situation, brought about through manufacturers increased production from aircraft ordered in the boom years of the 1980s, coinciding precisely with the massive downturn of air travel in the early '90s.

The worldwide recession which has hit the airline industry over the last few years, coupled with many hundreds of 'first generation' jets now reaching the end of their useful life has led to the unprecedented sight of millions of dollars worth of commercial transports sitting quietly in the desert, which during 1992 peaked at over 1,000 'desert airliners'.

By the summer of 1993 these numbers had slowly dwindled to a little over 750, as new markets were found for some of the older types and re-structured airlines took delivery of their newer aircraft. Sadly, well over half of those that remain in store come under the FAR Stage 2 and ICAO Chapter 2 noise limit regulations, making them far less attractive to operators than the newer, quieter types which are as readily available from the desert parking lots. Without the fitting of engine hush-kits to bring the aircraft into line on noise legislation, an expensive option for a new operator, these older types will almost certainly be condemned to a permanent retirement in the desert.

Arizona has long been known for its population of stored military aircraft at the huge Aerospace Maintenance & Regeneration Center (AMARC) at Davis-Monthan AFB, but other sites in the state are now acknowledged for their noticeable numbers of civilian types. Nearby Tucson International Airport has become a temporary home for numerous retired and transient airliners, generally smaller medium-range aircraft, but increasing numbers of larger jet types are now being handled. The town of Kingman, located at the junction of Interstate 40 and US Route 66 in the north-west of Arizona, became famous after the war for its enormous surplus aircraft disposal site. The sleepy airfield, known by its official title of Storage Depot Number 41, received well over 6,000 military aircraft in the immediate post-war years, as war weary veterans, together with a few hundred newly-built types were retired to the desert, and plans drawn up for their disposal. In recent years the field has taken on a look of its former self, though in much-reduced numbers, as redundant airliners from collapsed operators are parked up by their owners to await an upturn in their fortunes.

Pinal Air Park, situated some 40 miles north of Tucson at Marana, is the largest airliner storage site in Arizona. This vast, remote 2,300-acre site, once a CIA-operated air base, is the property of Evergreen International and home to their huge aircraft repair and maintenance facility. The number of stored airliners is still on the increase, due in no small way to certain operators, implementing specific fleet-storage deals, such as the US Air F28s and 727s, and Northwest Airlines 747s now arriving here. Shimmering through the heat haze, the lines of inactive airliners clearly visible from the nearby Interstate serves as a stark reminder of the present overcapacity in the airline industry, both in the USA and worldwide.

The business of airliner storage is competitive, and Mojave Airport in California is recognized as being the main contender for Marana in numbers, and variety of types presently in store. As well as being home to a number of civilian-run flight test companies, the airport boasts ample parking space and perfect weather conditions suited to the long-term storage of aircraft. During late 1993 well over 100 dormant airliners were parked up on the desert scrubland, as the more active residents of this friendly airport took full advantage of the clear skies and uninhabited locale in which to carry out their flight testing.

Las Vegas-McCarran Airport is the third 'major' airliner park in the south-western USA, and though not up to the enormity of either Marana or Mojave is still host to a growing number of airline fleet retirees, mostly of the older 'Stage 2' types such as the Boeing 727 and 737. The small airfield of Henderson-Sky Harbor, less than five miles from McCarran has also seen a number of retired aircraft arrive for temporary storage in recent years.

As ever more stringent environmental pressures on aircraft noise and emissions are introduced, so increasing numbers of the older jet types will eventually be retired, in favour of more 'friendly' alternatives. Added to this, the present 'glut' of newer more modern types available to existing operators, and new start-up carriers, mean the number of stored aircraft is unlikely to diminish significantly during the coming years.

Even with a recovery in the fortunes of the world airline business, America's ghostly 'Desert Airliners' look set to haunt the industry for the foreseeable future.

Graham Robson
Lincoln, England
February 1994

The incredible sight of hundreds of civilian airliners stored . . . lifeless, in the desert, now makes the news as much for the spectacle and curiosity it creates as for the recession that spawned the phenomenon. Four out of every five of the world's stored aircraft are parked up in the south-western United States of America, and the airliner proportion of this figure has grown quite significantly over the recent past, peaking at over 1,000 airframes during 1992. Part of the main storage area at Mojave Airport, California, contains this neat line of eighteen British Aerospace 146 regional jets last flown by US Air, which were inherited by the airline upon its merger with Pacific Southwest Airlines during the spring of 1986.

One of America's prominent operators, TransWorld Airlines, once under the financial control of the famous Howard Hughes (when it was known by the name of Transcontinental & Western Airlines), has not escaped the ravages of the recession in recent years. Saved from bankruptcy in early 1993, a reorganisation plan by the airline involved a series of fleet trimming measures which included the refinancing of some of its aircraft by selling and immediately leasing back into operation. Lockheed L-1011 TriStar N31011, bought new from the manufacturer in June 1973, flew its last service for the airline between St Louis and Kansas City on 24 October 1992, after which it was withdrawn from use and returned to its lessor, having flown over 53,000 hours for the airline in almost twenty years. Moving on to the Pinal Air Park, Marana, on 29 December it is seen heading up a line of three of the type.

This view could almost be the peak time rush-hour at one of America's major international airports were it not for the fact that all of these commercial widebodies are dormant and protected from the gruelling temperatures of the desert, in storage. This impressive line-up at Mojave in October 1992 includes an A-300 Airbus of Continental Airlines, Boeing 747s of KLM-Royal Dutch Airlines and America West together with an ex-British Airways L-1011 TriStar and Jugoslav Airlines DC-10-30. The larger types may stand a better chance than most of returning to service, as smaller airlines grasp the opportunity to expand their horizons and join the league of high capacity or trans-Atlantic charter operators, helped by the present glut of available equipment.

The Evergreen Air Center at Pinal Air Park has long been used to a full apron by airliners of all types and sizes, being home to the company's huge commercial aircraft repair and service centre. However, during the last few years more and more of the movements at this remote Arizona airfield have been inbound only as the numbers of stored aircraft steadily grow. Boeing 747-128, c/n 21141, flew commercially with Air France as F-BPVQ until 31 October 1991 when it was ferried across the Atlantic to Boston from Paris and registered to the Connecticut National Bank as N174GM, before flying on to Marana for storage. Accompanying it on the ramp in the picture is an ex-Lufthansa series 230B, and one of the two Boeing 747s flown by NASA as Shuttle transporters, which are maintained by Evergreen at Marana.

9

In the early 1980s Marana was in its infancy as an airliner storage site, numbers of aircraft being significantly lower than in recent times; which were mostly older types that became redundant through fleet renewals. Alitalia retired the remainder of its DC-8 fleet *en-masse* to Marana in early 1981, as they were superseded on the carrier's medium range routes by more modern types. The picture shows series 62H I-DIWH, in the company of five other examples at Marana in September 1981, all awaiting buyers. Purchased by the Peruvian national flag carrier Aero Peru, it took up the registration OB-R1249, was delivered in May the following year, and flew for this operator for the next nine years.

Houston-based Continental Airlines is one of the many major US carriers to have filed for chapter 11 bankruptcy protection in the early 1990s, brought about by the harsh recession in the airline industry and the rising fuel and operational costs of its fleet. A successful recovery package was negotiated with various cash investors which has helped to bring the airline onto a more secure footing. Substantial numbers of the Continental Airlines Boeing 727 and 737 fleet, which had been sold to and leased back from finance corporations, were retired from service and returned to their owners during the period of financial uncertainty, as illustrated by this collection of Boeing tri-jets at Mojave during 1992, with all markings painted out. As ever more stringent noise regulations for aircraft are enforced within the United States the number of retirements of this popular type will surely increase, with hush-kitting being the only viable, if expensive, alternative.

As part of Continental Airlines' new image, a completely revised colour scheme was designed, giving a stronger image to the airline's status as an international carrier, which is slowly being applied to the fleet as time permits. Airbus A-300 N234EA was built for Eastern Airlines, entering service with them in December 1983. It was bought by Electra Aviation Ltd on 23 March 1990 and was immediately leased back to Eastern the same day, eventually returning to Electra in March 1991 following the demise of Eastern. Continental Airline took up a lease of the aircraft on 1 April 1991, and it is seen in temporary storage at Mojave in October of that year wearing the fleet number #976 alongside the rear cabin door. It was re-registered into the Continental fleet sequence as N14976 in March 1992.

The two Boeing 727s seen in storage at Waco-Connaly Field, Texas, in October 1992 are ex-Pan American "Clipper Troubadour" N385PA, withdrawn from service in December 1991, and the ex-American Airlines N1908, a series 23 model which was retired to Waco on 25 September. The latter served American Airlines for almost twenty-seven years, with a short spell on lease to Global International Airlines during most of 1983, before finally being withdrawn from service on 15 September 1992, following a flight from Des Moines, Iowa, to Dallas-Fort Worth, with a total airframe time of 66,900 hours and over 50,000 landings. The distinctive American Airlines scheme has been changed slightly to a twin blue cheatline, following return to owner Aviation Leasing. It was flown to Miami for continued storage on 24 October 1992.

Not all of the airliners stored in the desert are there because of dwindling fortunes in the travel industry; a significant proportion are actually "trade-ins" as a result of fleet modernisation programmes. Happily, British Airways is one of the more fortunate operators, enjoying rather more success in these recessionary times due in no small way to some harsh cost-cutting a few years ago, which is now showing substantial benefits. The airline currently serves over 160 destinations from its two London operating bases, Heathrow and Gatwick; plans to break into the lucrative US domestic market were thwarted in 1992, when the US government refused to approve its plan to acquire a $570 million share in US Air.

G-BHBO is a TriStar series 200, designed for operation at "hot and high" destinations, and was delivered new to the airline in April 1981 and given the fleet name "The Morning Jewel Rose", later changed to "St Magnus Bay". It flew its last service for British Airways, from Lagos to Gatwick, on 21 March 1991 and was ferried to Mojave for storage on 17 June.

Eastern Airlines, once one of the largest and most prestigious airlines in the United States, became a casualty of the severe downturn in the fortunes of many US operators in the late 1980s. The Miami-based airline filed for chapter 11 bankruptcy proceedings in early March 1989, with limited operations continuing as the company strived to reduce its debts and cut costs. Its lucrative South American routes were sold to rival American Airlines, with various major domestic routes going to both US Air and Delta, who also obtained Eastern's Atlanta route "hub" in the deal. Sadly, the end came for Eastern Airlines at midnight on 18 January 1991 when all operations were officially ceased. Photographed at Marana, Arizona, in October 1992 was Boeing 757-225 N520EA, which at the time was coming out of storage and being prepared for flight once more. The aircraft left its temporary home of almost eighteen months on 29 October 1992, for a short flight to Phoenix where it was stripped of its existing scheme and was due to join the ranks of a more fortunate carrier.

TransOcean Airways was a North American charter operator based in Louisiana, originally flying only US domestic services under the name of Gulf Air (USA). Four Douglas DC-8 series 63s were acquired from Scanair in the spring of 1989 with plans to operate from New York with these long haul types to destinations in Europe and North Africa. However, problems arose when the airline was barred from using the name Gulf Air on international flights due to it conflicting with the operator in the United Arab Emirates. Plans to use the new name Sun International Airlines also ran into difficulty, with this title also being unsuitable. Finally, TransOcean was decided upon and the fleet was suitably repainted to show this. Operations under the new name were short-lived. The company filed for bankruptcy protection in December 1989, before finally ceasing all operations early the following year. Boeing 727-247 N2818W was originally a part of Western Airlines' fleet before being bought by Gulf Air in January 1985. It was photographed in storage on the remote side of Las Vegas-McCarran Airport along with a stable-mate in October 1991.

The almost meteoric rise of Midway Airlines in the 1980s seemed too good to be true. Flying routes out of their Chicago-Midway airport base, the airline soon began to rival many long-established US domestic carriers, and as passenger figures increased so did their route network. The company opened operating bases in both Miami and Pittsburgh to offer connections to most destinations along the eastern United States. Unfortunately, the airline could not keep pace with this over-expansion. Midway went bankrupt in March 1991, but continued to operate under the protection from creditors through chapter 11 regulations. A deal with Northwest Airlines to buy them out of their debt failed to come about and the airline went into liquidation in November 1991. In the ensuing months, their extensive fleet of DC-9 and MD-80 aircraft could be seen parked up at most of the desert storage sites. DC-9-31 N977ML was originally an Eastern Airlines aircraft, and is shown in the company of others at Marana.

Famous for being the most successful civil airliner ever, Boeing continue to build the 737 after more than twenty-six years in production. Despite the popularity of the type with over 2,500 built, the 737 is not the most numerous airliner of its class in storage, with slightly more than 100 examples parked up as of summer 1993. Of this figure, well over 90 per cent are older series 100 and 200 and Pratt & Whitney JT-8D powered versions, which are classed as "stage 2" jets under FAA and European noise regulations. Whilst hush-kits are available for these versions to bring them within current guidelines, the present glut of newer aircraft on the second-hand market will probably make this a rather expensive option for most operators. As a result new markets are having to be sought for the excess of noisy types presently stored in the desert. Braniff Airways and Continental Airlines B.737s are amongst those stored at Mojave, the former due to operator insolvency and the latter a result of the restructured carrier taking advantage of newer, more efficient types presently available from leasing companies. N4502W is one of seven ex-Braniff B.737s which were withdrawn from service, returned to Polaris Aircraft Leasing Corporation, and stored in April 1990.

The colourful markings of Evergreen International are a common sight all over the world on their mixed fleet of Boeing 747 and Douglas DC-8 aircraft, operating freight and parcel flights on behalf of United Parcel Service as well as their own services. The American supplemental cargo carrier based at McMinnville, Oregon, which grew from very humble beginnings in the helicopter logging business operating in north-western USA, also owns a large and mixed fleet of smaller freighter aircraft, including Boeing 727s and Douglas DC-9s. Following the loss of a freight contract for the US Postal Service in 1992, a substantial portion of the B-727 freighter fleet became temporarily redundant and were flown to the Evergreen Air Center at Marana for storage, pending a decision on their fate. N727EV was originally built as a passenger airliner for Braniff Airways in 1967, transferring to Evergreen ownership in 1984 after a short spell with Transbrasil. It was converted to pure freighter configuration in October 1984, and was pictured in storage in October 1992.

The basic design may be over twenty years old, but with more than 1,000 of the type produced the Boeing 747 is still in a class of its own as a long haul airliner. It is because of this uniqueness that the 747 has remained so popular, with certain examples moving on to second and third tier operators as national flag carriers introduce the latest variants into service and move the older versions down the airline "chain" There are presently over fifty Boeing 747s laid up, inactive in the desert, including this example last flown by America West Airlines. The Phoenix, Arizona-based operator introduced the 'jumbo' into their fleet in late 1989, inaugurating services to Honolulu from both Phoenix and Las Vegas in November of that year. Like so many other US airlines, America West underwent major change in the early 1990s in an effort to remain in profit, which included deferred payment on some of their aircraft purchases and the sale of their Honolulu-Nagoya route. This move rendered the B.747 superfluous to fleet requirements, including N532AW which was withdrawn from service at Phoenix on 12 October 1991; and moved into storage at Mojave three days later, following its return to the registered owner, the Bank of America.

Probably the most flamboyant North American airline, famous for its multi-coloured fleet, Braniff Airways became a victim of the de-regulation in the airline industry in the United States in the early 1980s, going out of business in 1982. Backed by the Hyatt Hotel chain they returned in March 1984, but by September 1989 had filed for bankruptcy once again. Services continued at this time, but on a much-reduced scale with flights serving only one-third of the original destinations. Operations ceased in November of that year. The airline arose from the ashes yet again less than a year later — as Braniff International — operating leased Boeing 727s financed by former real estate owners, but suffered another failure a mere thirty-seven days after beginning services when the airline declared itself bankrupt on 7 August 1991. Four of its colourful B.727s could be seen in storage at Mojave in October 1992, including this aircraft wearing the blue version of Braniff's revised "Flying Colors" livery. N8856E was originally operated by Eastern Airlines; bought by Aeron Aviation Resources in July 1991 it served for a short while in the fleet of Pan American before their cessation and was leased to Braniff International Airlines, wearing the fleet number 407.

Although most of the major airliner storage locations are in the deserts of southern California and Arizona, one or two airfields in Texas are now beginning to fill with redundant airliners. Ex-Continental Airlines series 15 Douglas DC-9s are lined up on a vacant hardstand at Waco-Connaly Field in October 1992. Situated some ninety miles south of the Dallas/Fort Worth area, this huge and remote former air force base has more than ample space available for the storage of unwanted airliners, and is home to the Texas State Technical Institute, who themselves own a number of retired aircraft with which to hone the skills of their students in the aeronautical trades. Douglas DC-9-15 N652TX is one of a number of short-range variants returned by their last operator to the financiers and lease companies which have helped to restructure numerous carriers in difficulties. These high-time airframes may now prove to be a liability to their owners as the major airlines swap older examples of their fleet for those with a fresher maintenance status, leaving the leasing company with expensive servicing bills for keeping the aircraft fit for a future operator.

Another "resident" at Waco, Texas, in the autumn of 1992 was this Douglas DC-10 series 10 G-BJZD, wearing the distinctive scheme of its last owner, UK charter airline Novair. The company was originally known by the clumsy name of British Caledonian (Charter) Ltd, being, as the name suggests, part of BCAL prior to that airline being taken over by British Airways. A name change to Cal-Air International occurred in 1985 before it was finally renamed Novair in December 1988. This particular aircraft was ordered and flown originally by Laker Airways, operating on their famous "Skytrain" services as G-GFAL, named "Northern Bell". Repossessed by the financiers on the demise of Laker, it was leased to British Caledonian in February 1982. Sadly, Novair suffered a similar fate, being put up for sale by its parent company, The Rank Organisation, in April 1990. The airline went out of business the following month.

Probably the most famous and well-known of all American commercial carriers, Pan American World Airways became a huge influence on worldwide air travel during its lifetime. The carrier was responsible for such aeronautical milestones as the introduction of jet transport to the United States, as the launch customer for both the Boeing 707 and Douglas DC-8, and the advent of high-capacity air travel with the introduction of the "jumbo jet" Boeing 747 in 1970. Thanks to this, its simple but distinctive colour scheme and famous "Clipper" names became a regular sight all around the world. In reality, the once immortal Pan Am could not stave off the inevitable. In a hostile economic climate and with mounting debts it continued to operate for a short while after its bankruptcy declaration of early January 1991, but the end came in December of that year with the spoils of its empire being claimed by its more fortunate rivals. These Boeing 727s stored at Mojave in October 1992 – including N8836E, the 818th 727 off the line – were all once the property of another failed operator, Eastern Airlines, and had been used on the cut-price and highly competitive east coast shuttle routes.

19

The stylised "M" and attractive red, white and grey trim aircraft of Chicago-based Midway Airlines are now resigned to the barren scrubland of the ever-growing desert aircraft parking lots of the USA. With newer and more environmentally friendly alternatives available, these older, noisier types will find it difficult to find work in the coming years in the United States, and will most likely find their way into the fleets of airlines operating in the less noise-conscious markets of South America and Asia. Parked out at Marana in October 1992, the DC-9 series 31 N933ML is typical of the many aircraft presently haunting the American airline industry, which has more aircraft than passengers.

The Canadian charter airline CP Air once possessed one of the most colourful liveries to be seen on any airliner, with its distinctive orange and red markings and polished metal undersides a common sight at many major European and worldwide destinations in the 1970s and 1980s. During 1986, however, the bright markings gave way to a more subtle midnight blue, red and cream scheme and the title was amended to Canadian Airlines. The operator can trace its history back to the early 1940s. Canadian Pacific was formed when the railway company of the same name acquired a number of small bush air operators. Since then the airline has absorbed a number of other carriers into its ranks, including Eastern Provincial in 1984 and Wardair in 1989. "Empress of Ontario" a Douglas DC-8 series 63 C-FCPQ was delivered new to the airline from Long Beach on 24 February 1968 and flew as one of CP Air's flagships for the next fourteen years, interrupted by a brief spell on lease to freight operator Flying Tiger Line in 1968. Relegated to second-line routes in the early 1980s following the introduction of DC-10s to the fleet, it was retired in 1982 and ferried to Las Vegas-McCarran airport for storage, where it was photographed in October of that year. Bought soon afterwards, the DC-8 returned to Canada flying for new operator Worldways Airlines, before moving on to a life as a pure freighter with a number of different American companies.

The BAC 1-11 design was typical of so many early types intended for use on high-capacity, short-range routes in the early 1960s as many leading airlines were entering the jet age. Making its first flight in August 1963, the type has enjoyed considerable success in its home country, but sales in the lucrative US market were disappointingly low due to considerable delays in obtaining the necessary FAA Type Approval, and with the growing number of aircraft types in the same payload/range category the BAC 1-11 never quite reached its potential. Seen stored at Las Vegas-McCarran Field in October 1990 in the company of numerous others, BAC 1-11 N1113J shows signs of its previous owner US Air, following its retirement from service in early 1989. As is now becoming common practice within the competitive airline business, these BAC 1-11s were traded back to another aircraft manufacturer as part of a deal for new equipment – in this case MD-82s – leaving McDonnell Douglas with the task of finding new owners for them. As the BAC 1-11 falls well within the class of so-called "stage 2 aircraft" (referring to those types that only meet the FAR and ICAO stage 2 noise limits) the prospective markets for such types are becoming limited. Nigeria, however, is one country without such regulations and has taken to the BAC 1-11 with enthusiasm, as many well-maintained examples are now available at very keen rates. N1113J now operates for Nigerian scheduled operator Kabo Air, from its Kano Airport base wearing the registration 5N-KBS.

A result of the de-regulation of the airline industry in North America was the sudden increase of small carriers, operating second-hand equipment and offering highly discounted fares on many services in direct competition with established airlines. The net effect of this was that monumental losses were incurred all round as prices tumbled in the fierce competition that ensued. As a consequence, many "upstart" carriers failed, sometimes within months, and some aircraft were destined to spend the following decade or so passing back and forth between lease companies and new operators. One such aircraft was this Boeing 737-219 N321XV, which was photographed withdrawn from use at Las Vegas-McCarran in October 1991. A fairly old example, delivered originally to New Zealand National Airways Corporation as ZK-NAC in August 1968, it operated in this region for the next eighteen years before being sold to CG Air Leasing Inc. in March 1986, when the American registration was allocated. Sold to another leasing company the following month the Boeing entered service with Presidential Airways in April 1986 and flew for less than a year with this carrier before being sold to Aviation Sales Company Inc. for onward lease to Phoenix-based America West. Returned a mere four months later, N321XV moved on to Panamanian operator COPA in October, returning from this short lease in early 1988. A further period of lease then followed, flying for Inter Canadian, before moving on to Greek carrier Olympic Airways for a twelve-month lease. Following return from this last period of operation for which the aircraft had received full Olympic Airways scheme, it was flown to Las Vegas for storage and, having had all Olympic titles and markings removed, it awaited its next move. Sadly, this was to be its last, as the B.737 was flown to Mojave soon afterwards to be sold for parts, and reclamation had begun by summer 1993.

The sight of so many BAe 146 regional jets laid up in the desert represents bad news for the British civil aviation industry. Of a total of over 200 146s sold so far, a staggering 25 per cent are presently lying idle in storage. This fact is due almost entirely to fleet disposals by certain carriers, retiring their whole complement of this type *en-masse*. This has the effect of severely depressing the market value of the type, both new and second-hand. These BAe 146s were previously operated by US Air, who retired the type from service in mid-1991, claiming that the aircraft was suffering excessively high operating costs within their fleet. N178US sits at the end of a long line at Mojave in 1992, with the high temperature of the summer sun beginning to take its toll on the paintwork.

US Air were operators of a fleet of British Aerospace 146 aircraft, which they inherited through the merger with San Diego-based Pacific Southwest Airlines in early 1988. N174US was photographed at Mojave in October 1991, parked up alongside the remaining seventeen examples in the US Air fleet, still wearing the three-tone red cheatline of Allegheny Airlines, by which name the carrier was known until 1979. Evidently some parts swapping had been going on among the fleet, judging by the US Air scheme rear door on this aircraft. Unlike the many older Boeing and Douglas types which abound at this remote Californian airport, the 146s are certain to be returned to commercial service before too long offering, as they do, an economical and acceptable alternative to the older types on offer in the used airliner market at present.

Heavyweights lined up at Pinal Air Park, Marana, awaiting an upturn of fortunes in the airline world. Boeing products dominate this view of the main ramp at the Evergreen Air Center, Marana, where work is carried out on the airframes to prepare them for an indeterminate time in storage. This process will include draining the aircraft's fuel systems and flushing through with light oil to protect the tanks, sealing up and covering with heat protective covers, all windows, doors, engines and open areas, and in some instances the complete covering of the tyres as a further measure against deterioration in the harsh sun, as seen on ex-Eastern Airlines Boeing 757 N520EA. The level of servicing and inspection the aircraft will receive whilst in storage is mostly dependent on the contract taken out by each specific customer, with higher value types which are more likely to be returned to service at shorter notice needing to be available to prospective customers as quickly as possible. Necessary maintenance costs to make the aircraft saleable only add to the burden of the aircraft financiers.

Boeing 727-225 N8876Z shows certain signs of its previous owner Eastern Airlines, with a blue nose cone and obvious marks where the cheatline has recently been scrubbed off. The aircraft was repossessed by its registered owners, General Motors Aircraft Credit Company, when Eastern Airlines collapsed in early 1991, the ownership changing once again in early 1992 when it was registered to Household Commercial Financial Services. The lack of markings on this particular aircraft, photographed at Marana, in October 1992, could signify better times ahead for it; possibly meaning that it is being prepared for transfer to another, hopefully more successful, operator.

The smart scheme worn by this series 30C Boeing 727 of Evergreen International belies the fact that the aircraft is now well over twenty-five years old. Originally delivered to German flag carrier Lufthansa as D-ABIO, it has flown since then as PP-SRZ for VASP in Brazil and with Air Tungaru, who operated it as T3-ATB on their Honolulu service from its base at Tarawa on one of the Gilbert Islands in the Pacific Ocean. N726EV is actually the third American registration allocated to the aircraft during its time with Evergreen. One of a number of Evergreen International B.727s presently stored at their Marana base north of Tucson, this example is unlikely to fly again after slight structural corrosion was discovered in its tail. Since then it has given up numerous parts to help keep others flying.

Even though the name Braniff has, once again, disappeared from airport timetables, it is not beyond possibility that in future years it will return; the airline has, after all, been resurrected on at least three occasions in the past. Photographed at Tucson International Airport in late 1992, Boeing 727-214 N409BN is sealed up and tied down in the corner of Hamilton Aviation's maintenance area. This twenty-four year old airliner had spent time operating for Denver-based Frontier Airlines before becoming the property of Braniff in 1971, and has been an active member of numerous commercial companies since 1982.

The general over-capacity in the airline industry is no more evident than at Mojave, where over 10 per cent of all the world's stored aircraft are parked. Familiar names and colour schemes abound at this vast yet relatively quiet airfield, situated in Antelope Valley to the east of the Tehachapi mountains north of Los Angeles. The business of storing airliners has grown in recent years, with some sites beginning to take on a similar look to the huge post-war military surplus sites, which abounded in the same areas. Most numerous, typically, are the dominant fleet types such as the Boeing 727, 737 and, as illustrated here, Douglas DC-9, whose numbers in storage represent around 17 per cent, 9 per cent and 15 per cent respectively of the world totals for these types. Fenced in and parked ever more closer as numbers increase, this picture shows examples from two failed carriers, Eastern Airlines and Midway, together with those of Delta Airlines. The latter operator, in a shrewd deal, recently traded back twenty-six examples to McDonnell Douglas as part of a purchase agreement on newly-built MD-11 widebodies, which has left the manufacturer with a stock of old airliners to dispose of as well as new equipment to sell. Several of these "pre-owned" DC-9s have since found gainful employment in Mexico and several South American countries.

Part of the collection of millions of dollars worth of commercial aircraft that greet the unsuspecting passing traveller or casual visitor to the sleepy Californian town of Mojave. This view illustrates the western end of the airfield, adjacent to the town, and includes no less than eighteen BAe 146s, seventeen Boeing 727s and thirteen Douglas DC-9s. At its peak in 1991 the airfield was home to almost 170 airliners, all looking for work in the recession-hit airline industry. It is a sad fact that the severe downturn in the financial fortunes of most airlines coincided almost exactly with the aircraft manufacturers' peak production rates, a result of the many new airframes ordered during the boom years of the 1980s, which has led to massive over-capacity. Consequently, many operators chose to park up their older "second generation" equipment in favour of the newer equipment, which helped to swell the numbers of retired airliners to well over 1,000 in 1992.

The impressive performance of the BAe 146, and its ability to operate in smaller regional airports with strict noise abatement procedures should ensure that better times are ahead for these ex-US Air aircraft stored at Mojave. Unfortunately, the same cannot be said of the ex-Continental Boeing 727 parked close by. With older and noisier engines and higher running costs, partly brought about by the need for a third flight-crew member on these tri-jets, the Boeing 727 could even lose out to its equally abundant rival, the DC-9.

With company headquarters now at Arlington, Virginia, US Air began life as All American Aviation in the late 1930s, becoming Allegheny and establishing an operating base in Washington DC. Its route network linked more than 100 locations, and a change of image was provided by a new colour scheme, prior to the name-change to US Air in October 1979. Two US domestic operators were absorbed into the airline, Pacific Southwest during April 1988 and Piedmont in August 1989. It was through the first of these mergers that the airline acquired its fleet of BAe 146s. As the photograph depicts, not all of the aircraft managed to attain the full US Air scheme prior to being withdrawn, with five of the assembled eighteen at Mojave still in basic Allegheny colours.

The registration N8944E on this Douglas DC-9 series 31 betrays the fact that it was once in the fleet of Miami-based Eastern Airlines, and has therefore, sadly, been in a similar situation once before. Its present owner, a lease company, bought it from the bank which repossessed the aircraft on Eastern's demise and leased it to Midway Airlines in July 1991, during their ascendancy. A short-lived tenure with the Chicago-based operator saw it return in November to join the lines of similar types at Mojave.

Marana, to the north of Tucson, is presently seeing significant growth in the quantities of stored airliners, in particular the larger types, with numbers of 747s as well as both types of long-range tri-jets present during 1993, as the photograph depicts. German flag carrier Lufthansa began to retire their Douglas DC-10-30s in early 1992, in anticipation of the introduction of Airbus Industries' new long-hauler, the A-340. D-ADGO heads a line of five DC-10s in October 1993, which includes two other ex-Lufthansa aircraft, one from Brazilian carrier VASP and an all-white example, with ex-Air Canada TriStars in the far background.

The appearance of the desert storage parks is prone to change quite dramatically at short notice, not least with the arrival of significant numbers of large widebody types such as the Boeing 747s Mojave has received in recent months. A specific deal with Northwest Airlines will see a dozen examples of the airline's early series 100 models arrive for storage, pending sale, as the latest "high tech" 747-400s are introduced into the carrier's route network. Resplendent in Northwest's latest colour scheme, N604US is seen at Mojave in October 1993, parked between a stable-mate in old colours and an ex-Canadian charter operator (Nationair) example. N604US was bought new from Boeing in June 1970, in the days when airlines actually owned their aircraft, and spent its whole life flying for the airline from their Minneapolis-St Paul base.

The barren Tehachapi mountains, which tower to a height of more than 6,000 feet and separate the arid Mojave desert from the greener, Central Valley region of California, seem to dwarf Boeing 747-206B PH-BUG, parked on the north side of Mojave airport's secondary, east-west runway. Conditions at this remote airport are eminently suitable for the long-term storage of aeroplanes, with extremely low annual rainfall, almost no humidity and the hard sun-baked ground ideal for parking. The distinctive colours worn by this 747 readily identify it as having been part of the Dutch airline KLM. This particular example had latterly been flown by the Indonesian state airline Garuda, on lease from KLM, returning to Amsterdam-Schipol in August 1991. Withdrawn from service upon its return, it flew to Mojave in December of that year to join the ranks of the desert airliners.

The slightly incongruous sight of seven members of British Airways' long-haul fleet parked up, dormant, in the Californian desert makes quite a picture. The photograph illustrates well the sad irony of the situation when one realises that the view represents the past and future of the airline's medium and long-range fleet plans. The Lockheed TriStars in the foreground, three series 200 "hot and high models" and two converted series 50 versions, have all been traded in to Boeing as part of the airline's B.767-336ER purchase deal. The factory-fresh Boeing 767s, which were delivered direct from the manufacturer's Seattle production plant are devoid of titles and markings; and will replace the Lockheed tri-jets on European, Middle East and North American routes, and were only temporary residents of Mojave when photographed in October 1991.

Air Canada was the third purchaser of the new Lockheed L-1011 TriStar in May 1973, and the first non-American operator of the type, behind launch customers Eastern Airlines and TWA. The Canadian flag carrier has now phased the type out of its fleet, preferring to concentrate mainly on Boeing products for its long-haul services. The Pinal Air Park, Marana, has now become home to a high number of the ex-Canadian aircraft which have still to find new customers. This collection of four TriStars all completed their last services for Air Canada in October 1990, and were initially stored at Montreal-Mirabel Airport, pending a firm decision on their future. Within a month they had been ferried south to Marana, and were sealed from the elements with all titles removed.

Even with acres of desert filled with new and used airliners of all sizes and categories, certain types are expected to fare better than others in future second-hand world markets. The Douglas DC-10 is generally expected to follow the path of its manufacturer's predecessor, the DC-8, as being a popular type for freight conversion, which could significantly prolong the life of certain examples. Significant numbers of the Douglas DC-10 are now available following the mass retirement of more than thirty examples by American Airlines into storage recently, which is bound to have some effect on the second-hand value of the type. This former Jugoslav Airlines DC-10-30 YU-AMC had been flown to Mojave by October 1992 following return to its lessor. The one-time Swissair aircraft, of 1973 vintage, was re-registered as VR-BMP by February 1993 in preparation for a further period of lease with another operator.

Pan American Airlines was the launch carrier for the unique Boeing 747, and by the time it entered service in 1970 the manufacturer had secured orders for some 190 747s from twenty-eight airlines worldwide, such was the impact this enormous type created. It is a sad fact that Pan Am will not be around to celebrate the type's twenty-fifth anniversary next year, having succumbed to the ravages of recession, competition and more than its fair share of bad luck. A depressing reminder of Pan Am's 747 days is depicted here, as the one-time N657PA (the 129th example to be produced) shows the last vestiges of a former scheme (Marana, late 1992). Having had the paint from its last operator, Evergreen International, stripped from the airframe, the rudder from another ex-Pan Am 747 which was being broken up alongside has been fitted to the aircraft temporarily.

The smart yet simple colour scheme of German flag carrier Lufthansa is easy to recognise on this Douglas DC-10 series 30, photographed on the main ramp at Marana in October 1992. The airline had begun to replace its DC-10s with an increased Boeing 747 fleet, and by mid-1990 at least one DC-10 was withdrawn from use at Frankfurt, though it had returned to service by early 1992, only to be retired to the desert some months later. The fleet name "Leverkusen" can still be seen on the forward fuselage, with the airline's flying bird motif still very evident on both the nose and wing engine, even though the paint has been removed. D-ADAO was delivered new from the Long Beach production line of McDonnell Douglas on 12 November 1973, and has spent its entire career, to date, flying for the same airline.

In order that these aircraft might endure the harsh conditions which the hot deserts of the south-western United States create, a special preservation process has evolved over the years to help maintain the airframes in the highest possible condition. The TriStars, pictured here at Mojave, illustrate well how all intakes and openings on the airframe are blanked off to prevent dust or water getting in, with all the flight-deck and cabin windows sealed and covered in heat-reflective foil, which has the double benefit of protecting the glass areas from the sandblasting-effect strong winds can have and of helping to maintain a fairly constant temperature within the aircraft cabin. These protective measures may well have been in vain for these ex-British Airways aircraft; G-BHBN and colleagues are now the property of Boeing, having been traded as part of the carrier's B.767 purchase, and will probably be dismantled for parts. This option could give the owners a better return on their investment, and would also take out of the market a handful of Lockheed aircraft which could potentially be sold against Boeing products.

The long-held ambition of European aircraft manufacturers to compete against the almost complete domination of the USA in the world market of commercial aircraft came about with the Airbus A-300. First taking flight in October 1972, the Airbus was to be the spearhead in a concerted campaign to offer an airliner "family" alternative to Boeing and Douglas products. Initial high hopes seemed to be dampened by poor sales in its early years, but 1979 saw a major break into the American market when Eastern Airlines placed an order for thirty-four new airframes as well as buying the four examples it had been operating on evaluation. The type has since flown for such carriers as American Airlines, Pan American and Continental. Eastern Airlines N221EA sits patiently on the hard-baked desert floor at Marana, alongside a former B.727 fleetmate and a number of ex-Air Canada TriStars. This Airbus was part of the many "sale and lease back" deals carried out by the airline in the late 1980s in a bid to raise capital and maintain profitability. Sadly, Eastern ceased trading in January 1991 and the Airbus was ferried out to Arizona for storage.

Air Canada once operated a fleet of eighteen Lockheed Tristars, both the long-range series 500 variant and the original series 1 models, which were delivered in the early 1970s. American carrier Delta Airlines bought the long-range models during 1991/2, and the remaining series I aircraft in the fleet were all withdrawn from service by the end of 1990.

Fleet number 506, C-FTNF, operated a Toronto-Montreal service as its final commercial flight for the company on 26 October 1990, and was retired immediately afterwards with a total airframe time of 41,970 hours and 19,308 landings in its logbook. Ferried from Montreal to Marana on 23 November, it is now awaiting a buyer. A

total of thirty-eight TriStars are presently in storage, from a worldwide fleet of 240 aircraft, and recent plans to convert the type for cargo operations should see these numbers slowly dwindle in the coming years.

From a high point of over 1,000 airliners in storage in 1992, the airfields involved in the South-west USA are slowly seeing a decline in the numbers as factory-fresh airframes are finally delivered to their new owners, and markets are found for some of the more fortunate older types. Regrettably, a significant number of the older types are "stage 2" aircraft, unable to comply with noise legislation enforced by the US FAA and the European ECAC (European Civil Aviation Conference). These rules have

effectively closed the door to any fleet increases of these types in both regions, in the form of a stage 2 "non-addition to the register" rule, which prevents any foreign registered non-stage 3 type moving into these two markets, or between the two. The Douglas DC-9 is a prime example of this particular type of aircraft, available in sizeable quantities from the desert parking lots but without any real prospect of returning to service in its own country; customers are sought elsewhere, with most

interest stemming from South American and Asian operators. Another alternative available to the frustrated owners is to have the aircraft "parted out", broken up on site and its parts sold as spares to other operators. Ironically, a DC-9 or Boeing 727 is now worth, potentially, twice as much as spare parts as it is whole, realising anything up to $6 million once the cutters have set about the airframe.

A sight such as this spells bad news for the airline financiers and lease companies, who now represent the main owners of the majority of the commercial fleets of US carriers. Sold by the airlines to these third parties and leased back immediately, the aircraft continued to operate in their respective fleets, allowing the airlines to take full advantage of their inflated value at the time, the final responsibility for their disposal being transferred to the registered owners. The only real winners in the industry at present are the owners of the airfields on which the hundreds of airframes are parked. Rates for storage depend on the level of attention needed but generally run between $250 and $300 a month for narrow body types, and up to $450 a month for a 747, with a reduction for five or more airframes. Twenty-plus DC-9s, mostly from two of the biggest airline failures of recent times, Eastern and Midway Airlines, sit in the sun at Mojave in October 1991.

British Airways are the largest non-US carrier to operate the Lockheed TriStar, taking their first example in October 1974. A total of twenty-nine L-1011s were to fly for the airline over the next twenty years, though the type is now operated only on charter routes under the Caledonian division of the airline. G-BGBC last flew for British Airways on 31 October 1991 when it operated a Paris-Charles de Gaulle to Heathrow schedule. It was withdrawn from use on arrival at Heathrow, and had its company titles removed by 6 November pending delivery to the USA for storage. It was flown to Mojave to join the other ex-British Airways TriStars, routeing via Bangor, Maine on 6 January 1992 and is shown ten months later having had its three Rolls-Royce RB-211 engines removed.

Although most of the inmates of the desert storage sites are still resplendent in their last operator's livery, others prefer a degree of anonymity on retirement. However, certain schemes are sufficiently distinctive as to be recognisable even after repainting. Consequently, the spray shop cannot hide the fact that Boeing 737 N148AW was last flown by America West Airlines, to which fact the tail logo and fin tip attest, not to mention the registration suffix. America West is a major scheduled carrier based at Phoenix-Skyharbor International Airport, with other "hubs" at Las Vegas-McCarran Airport, Nevada and Columbus, Ohio. A product of the de-regulated US airline industry, it was formed in February 1981, but became another victim of the overcapacity and cost-cutting in the airline world, filing for chapter 11 bankruptcy protection in June 1991. A successful recovery plan was executed, involving cost-cutting and realignment of routes together with fleet and staff reductions with the intention of emerging from chapter 11 in early 1993. The 737 here, photographed at Marana in October 1992, had been returned to its registered owner Citicorp, a leasing company, on 25 September and had moved into storage two days later.

Probably unique amongst the hundreds of airliners presently stored is this Boeing 707-138B registered VR-CAN, which can technically be termed an executive jet. Last flown in October 1981 when it arrived at Marana, the Boeing was originally in the fleet of Australian national flag carrier Qantas as VH-EBH, having been delivered in July 1961. Following seven years of operation with Qantas the Boeing was sold to British West Indies Airways in September 1969, taking up the registration 9Y-TDC. It flew on the carrier's international routes from its Trinidad West Indies base for the following eight years, after which ownership changed and it became the property of aircraft brokers Omni International. Sold to Euro Financing soon afterwards, the aircraft spent a lot of its later life parked at Stansted Airport, Essex, and carried out very little flying. The Boeing is fitted out with a very plush executive interior, including gold fixtures and fittings in its bathroom, and was reportedly used by the late Shah of Iran during his years of exile. Since its arrival at Marana, the aircraft has remained parked in almost the same spot on the airfield, and although fees are paid for its parking, the level of attention the airframe receives would seem minimal judging by its rather unkempt appearance.

The "Desert Airlines" are not confined exclusively to jet operators, with the recession in the airline industry not discriminating between carriers. Following de-regulation of the airline industry in the USA in the late 1970s, the number of small regional carriers multiplied enormously, and not having the vast oversupply of second-hand jet equipment available as today, most were contented to begin with mid-size prop types. Golden Gate was typical of these airlines, progressing onto the de Havilland DHC-7 from smaller Fairchild Metroliners, but it ran into difficulties following over-expansion, coming after their acquisition of competitor Air Pacific and its aircraft. The carrier suspended operations at the end of August 1981, and its aircraft were withdrawn, pending sale. Nine of the carrier's DHC-7s were flown to Marana for storage in August 1981, including N705GG which had been delivered new to the airline only three months earlier. Bought a few months later by a finance company, the aircraft has since flown for a succession of third level carriers.

The distinctive colours worn by this pair of Convair 580s are those of Sierra Pacific Airlines, a charter carrier with a long association with the type. Originally a Bishop, California-based operator, Sierra Pacific took over the assets of Mountainwest of Tucson in late 1978, gaining six Convairs in the merger and transferring their base to Tucson. The airline continues to operate the Convair, though in recent years it has been leasing out aircraft to other airlines for operation as part of a major carrier's feeder services, often flying in the host airline's colours. Sierra Pacific used Marana as a base for their operations in Tucson, and these Convair 580s, N73112 and N73106, were both photographed at Marana in storage minus all titles and logos, having last flown during 1992. The airline has since entered the jet age, and is now flying a small number of Boeing 737s, which could eventually mean the end for the popular Convair.

As well as being home to a number of temporarily and permanently withdrawn airliners, Tucson International Airport is also home to Hamilton Aviation, renowned for their vast experience of all variants of Convair. This association with the type includes overhaul, repair and conversion, as well as acting as broker for the sale of the ever-popular airliner. Photographed on the Hamilton ramp in late 1989, this fine-looking Convair 580 TG-MYM wears the markings of Guatemalan scheduled operator Aeroquetzal, which flew services from Guatemala City to Cancun, Puerto Barrios and San Pedro Sula. This particular aircraft has an interesting history. Beginning life as Convair 340 (c/n 96) N8420H, owned by Union Producing Co, and delivered on 30 July 1953, the aircraft suffered a severe fire at Shreveport, Louisiana, which resulted in the fuselage being badly damaged. Returned to Convair, the aircraft was rebuilt in May 1956 as a model 440, and was given the new manufacturer's construction number 327A; re-emerging as N8444H. It was returned to Union Producing Co. Sold to US domestic operator North Central Airlines in March 1966, the Convair later joined the "conversion" production line of Pacific Aero at Burbank, California, on May 1969 and became an Allison turboprop-powered model 580. North Central merged with Republic Airlines in 1979 and disposed of the Convair in late 1983 following an accident, before it eventually ended up in Guatemala. The colourful Convair arrived at Tucson in September 1989 and was withdrawn from use, still present two years later.

Air Bridge Carriers, part of the Hunting Aircraft Group under which name they now operate, were responsible for introducing the classic Lockheed L-188 Electra to the British register in the summer of 1991. The type has since flourished, becoming the mainstay of their fleet, and also a number of others since that time. This particular example was photographed at Mojave in late 1991 in storage, following a series of noise and engine tests required for UK certification, a this airfield which is world famous as a civilian flight test centre. Acquired from NWT Air of Yellowknife, North Western Territories, where it flew as C-FNWY, the simple change of markings to G-FNWY was made for its flight to California on 23 May 1991. The lack of a freight door on this particular Electra rendered her slightly impractical in the fleet of a cargo carrier, and she was put up for sale by Air Bridge. A buyer was soon found for this immaculate aircraft, and it was ferried to Opa Locka, Florida, in early 1992 as N3209A and prepared for passenger operations with Indian Ocean Airlines based at Perth, Western Australia, where the markings VH-IOB had been allocated. Sadly, type certification problems in Australia forced the carrier to seek alternative equipment, and the Electra returned to the United States for operation within the fleet of JBQ Aviation as N351Q.

The attractive blue and yellow markings of Mexican charter carrier Aerolitoral are seen on one of their NAMC YS-11As, XA-ROV, photographed at Marana in storage during October 1991. Servicios Aereos Litoral is a subsidiary of flag carrier Aeromexico, acquired in November 1990, and flies feeder services to the main carrier's hub under Aeromexico flight numbers. XA-ROV once flew as part of the fleet of Simmons Airlines as N274P, operating as part of the American Eagle network, the commuter feeder carrier to American Airlines. Purchased by Short Bros (USA) in January 1987 as part of a fleet purchase deal with Simmons Airlines involving new Shorts Aircraft, the YS-11A was ferried to Tucson in March 1987 for a period of storage, during which time ownership changed twice, before joining the fleet of the Mexican airline in August 1989. It returned to Arizona once again for storage, and in May 1991 it was flown to Marana with two other fleetmates where it was allocated the American registration N4246K.

Last flown in its homeland as JA8640, operating for TOA Domestic (later renamed Japan Air Systems) this NAMC YS-11 was the sixth example of the type to roll off the production line in April 1965. During the late 1980s the YS-11 underwent a resurgence in the United States as numerous early production models were imported from Japan. Still in its TOA colours, N206VA was bought by aircraft broker Corns & Co. (Hong Kong) Ltd in early 1989, and was flown to the small Las Vegas-Skyharbor/Henderson Airport by November 1990 with another example. Both airframes were parked up on the edge of the airport's main ramp, and were subsequently stripped of most of their useful parts including the engines and windows for sale as spares.

The Convair 640 is a development of the successful model 440, fitted with two Rolls-Royce Dart turboprops. The conversion added much to the performance of the aircraft, putting the type in direct competition with the NAMC YS-11, Vickers Viscount and Fairchild F-27 for selection by the many domestic carriers keen to move on from piston power in their fleets. This particular example, C-FCWE, having been built as a model 340 and later converted to a 440, became the first 640 variant, out of a total of twenty-seven to be converted when it received a pair of Darts in June 1966. Latterly operated by Cleveland, Ohio-based Wright Airlines, the aircraft was later to star in the movie *Great Balls of Fire* painted to represent an American Airlines Convair of the 1950s. Returned to Tucson immediately afterwards, it was repainted to join the fleet of Canada West Air in the summer of 1989, as illustrated here. Sadly, the airline folded early the following year, and the Convair was re-registered N862FW for operation by Gambcrest, a Senegalese carrier, and was delivered via Luton in October 1990. Tragedy was to strike less than four months later when the Convair crashed in southern Senegal with the loss of thirty-nine lives.

Now out of production, the Shorts 330 was a much-refined development of their Skyvan design, which had a much greater appeal for third level operators around the world. Many North American carriers purchased the type, one being Command Airways which flew as part of the American Airlines commuter network system American Eagle in a variation of the major carrier's colour scheme. This Shorts 330 N344SB was part of the Command fleet until it was returned to Short Bros (USA) mid-1988 as part of a deal to purchase larger 360 models prior to the airline being merged with Nashville-based Flagship Airlines, another American Eagle carrier. The 330 was initially withdrawn and stored at Reading, Pennsylvania, before being ferried to Las Vegas-Skyharbor/Henderson by late 1990. Sealed up against the elements in October 1991, it was re-registered N76NF in December the following year for operation by Freedom Air of Guam.

Looking a little out of place in the desert is Grumman G-111 Albatross N118FB, one of a number of these ageing amphibians presently parked up at Marana, suitably protected against the hot harsh desert climate. Chalks International Airlines, the world's oldest operating airline, are part of the hotel and casino group Resorts International, flying a small fleet of Grumman Turbo Mallards between southern Florida and the Bahamas.

During the early 1980s Resorts International announced a programme to develop ex-military HU-16 Albatross airframes into 28-seater commuter amphibians to cater for the needs of their seaplane operations with Chalks International and associate company Antilles Air Boats. The conversions, costing over $3 million per aircraft, involved extensive modifications to meet stringent FAA certification requirements, which included the

fitting of two additional emergency exits and a certain amount of flight-deck modernisation. A total of twelve aircraft were converted, with the type emerging as the Grumman G-111 and entering service with Chalks in January 1982. This Albatross, N118FB, was once part of the Royal Canadian Air Force where it served as 9304, and has been parked at Marana since early 1990, held in reserve in the desert alongside a number of other similar types.

The orange and red livery of Wright Airlines, a once-busy third level carrier, is now but a distant memory, having joined the increasing ranks of airlines which have failed in the United States in recent years. Based at Cleveland, Ohio, Wright Airlines flew an extensive route network throughout the southern Great Lakes region and beyond. The airline was declared bankrupt in September 1984 and filed for chapter 11 protection which enabled operations to continue, though on a reduced scale, and although its fleet of Shorts 330s was repossessed by the manufacturer, services were maintained with its remaining Convair 600, 640 and EMB-110 Bandeirante aircraft. In early 1985 a deal was signed with a Puerto Rican travel company to provide commuter services from San Juan under the name Puerto Rico Express using three Convair 640s, commencing on 1 May that year. The venture was short-lived, however, and the Convairs were returned to their parent company soon afterwards. Convair specialists Hamilton Aviation, based at Tucson International Airport, took possession of Wright Airline's redundant Convairs when the airline finally folded. Parked up at Tucson, the aircraft were used as a source for valuable spares by Hamilton, who provide an extensive support and maintenance service to operators of this popular type. In this October 1991 photograph, Convair 600 N94224, still wearing the full colours of its previous owner, is parked outside the Hamilton compound at Tucson Airport with a number of similar types following the removal of its valuable Rolls-Royce Dart engines. The area around the flight-deck windows shows signs of the aircraft's protective seal having been removed, signifying its impending demise. This covering is applied for extended periods of storage, both to preserve the window glass and to maintain a reasonable internal temperature, avoiding possible damage to rubber parts and expensive functional components.

The Vickers Viscount became famous as the first type to offer scheduled commercial passenger services in a turbine-powered aircraft when it began flying for British European Airways on 29 July 1950. Developed considerably since that date, the type was an immediate success with airlines and passengers impressed by the type's comfort and performance. A crucial point in the Viscount's success was its breakthrough into the highly competitive North American market when Capital Airlines ordered forty model 745s in August 1954, and the type maintains the record as being Great Britain's highest-ever produced airliner. Even into the 1980s the Viscount could still be seen at many airports around the world, though its numbers were dwindling. In North America, however, one man did develop a certain liking for the type, accumulating the largest collection of Viscounts anywhere in the world, which peaked at thirty-four examples in 1983. Ron Clark of Tucson was the owner of The Go Group, a combination of affiliated companies founded in Burbank, California, in the early 1970s. Acclaimed for their superb furnishings and with the proximity of the entertainment industry headquarters at Hollywood, the Viscounts soon became popular with many famous names in the film and music business, who would soon become regular customers for the luxury and level of service of Go Transportation. In mid-1980, having outgrown its Burbank home, Go Transportation moved to Tucson International Airport, where the immense fleet of classic Viscounts became a well-known sight. This Viscount, series 814 N145RA, was one of a number which was used on the company's brief flirtation with scheduled operations in the early 1980s – under the name Royal American Airways – flying between Tucson, Las Vegas, Long Beach, San Diego and Lake Tahoe. Sadly this was a short-lived venture, and the aircraft returned to its more normal lifestyle of transportation for the rich and famous, as the name on the nose confirms. The aircraft had operated for Lufthansa, British Midland and Israeli airline Arkia before taking up American markings in June 1982, and is photographed at Tucson ten years later minus engines and in need of some attention, not having flown for a good number of years.

In order to maintain a fully airworthy collection of Viscounts, Go Transportation acquired a number of airframes from various sources to provide a healthy supply of engines and spare parts for the active members of the fleet. One such source was Arkia Israeli Airlines, a domestic scheduled passenger and cargo operator, the country's only user of the Viscounts, operating nine examples. Bought by Go Transportation in May 1982, 4X-AVE was delivered fully marked up in a later variation of its last owner's colour scheme. The aircraft never flew again, but continued to donate parts until its owners went out of business in the late 1980s. Seen here on a remote parking spot at Tucson in late 1992, suitably secured with a fifty-gallon oil drum attached to its nose undercarriage, the Viscount now looks a little the worse for wear.

Las Vegas-McCarran Airport is another of the large-scale aircraft parking and storage sites located in the south-western USA, situated close to the city's famous "strip", as the view behind this Douglas DC-3 N138D illustrates. This September 1981 photograph shows four DC-3s of Royal West Airways and their associate company Pacific National Airways. The carrier provided charter services to numerous locations in the southern California/Nevada region, including Lake Tahoe, Grand Canyon, Las Vegas and Burbank. Operations had ceased following the collapse of both companies in mid-1981, and the aircraft were flown to Las Vegas for storage, pending sale. As one would expect, the value of operational DC-3s in good condition helped to secure safe futures for all four examples, which had all moved on by the following year.

As well as accommodating hundreds of temporarily redundant airliners, Arizona is also home to a few operators of aircraft of more mature years which, by the nature of their specialised roles, spend a good deal of time on the ground. Chandler-Memorial Field, situated on the southern edge of the city of Phoenix, is a most unlikely place to find one of the largest collection of large, multi-engined piston-powered ex-airliners. The airfield, with one paved runway and a ramp area which is no more than cleared scrubland, is the operating base for T & G Aviation, a USFS contract fire tanker operator flying DC-6, DC-7 and ex-military C-130 Hercules, and Biegert Aviation, a company with a fleet of C-54 aircraft configured for aerial spraying work as well as freight contracts. T & G have long been associated with the highly specialised task of aerial fire suppression, and have accumulated a sizeable collection of ex-commercial DC-6 and DC-7 airframes, a significant proportion of which are configured for the fire-fighting role with a 3,000-US gallon capacity under-fuselage retardant tank. In February 1991 T & G acquired the aircraft of Macavia, another tanker outfit based at Santa Rosa, California, and added their DC-6 fleet to the assembled DC-7s at Chandler. Photographed here in October of that year at the head of a line of similar types, DC-6 N555SQ "Tanker 45" still wears the logo and titles of her former owner. This classic aircraft was delivered new to United Airlines in June 1957, and was converted to its present state in the early 1970s. Chandler has proved a very suitable location for such an operation not only because of its proximity to potential fire sites throughout the western States, but also because of the availability of aircraft and engine spares in this part of the USA, and the very helpful climate Arizona offers, minimising airframe corrosion in its very dry atmosphere.

Still wearing the red and blue colours of the Japanese domestic carrier Japan Air Systems (formerly TOA Domestic), with whom it flew as JA8648 until retirement in November 1991, is this NAMC YS-11 photographed at Tucson in October 1992. Cancelled from the Japanese register in November 1990 upon its sale to an aircraft broker, the aircraft was given the registration P4-YSE for operation with Air Aruba. However, twelve months later ownership changed once again, and the US marks N991CL were allocated. During early 1992 the YS-11 was flown to Tucson to be stored in the Hamilton Aviation compound, where it received a further registration change, once again taking up Aruban marks, this time P4-KFC. In this October 1992 photograph, the aircraft looks ready to fly, although it was still at Tucson twelve months later wearing its hastily applied Aruban registration.

Sharing the desolate airfield of Chandler-Memorial is Biegert Aviation, whose Douglas C-54 fleet came entirely from military surplus sales at Davis-Monthan Air Force base in nearby Tucson. Owned by Mr Max Biegert, who is also the proprietor of the famous Grand Canyon Railroad Company, a number of the C-54s have been fitted with a system of spray-bars on the wing trailing edge and are available for contract crop and pest control, though in recent years they have done little flying. One or two aircraft are configured for *ad hoc* freight charter work, though in reality spend more time parked up, inactive, at Chandler. N44904, the one-time US Navy Bu.No. 56530, shows small change from its military scheme, with the simple addition of a red cheatline.

SPARE PARTS, FILM STARS AND SCRAP

Without doubt, Davis-Monthan Air Force base on the outskirts of Tucson, Arizona, holds the largest concentration of aircraft in one place anywhere in the world. As home to the Aerospace Maintenance & Regeneration Center (AMARC) the base receives all US military aircraft withdrawn from service for storage, both short- and long-term, pending their reassignment either to an active unit or onward transfer to a friendly nation, or even their disposal as surplus to requirements. Since the early 1980s, however, AMARC has been the recipient of 186 former civilian Boeing 707 and 720 airliners, which were purchased by the military for use in an update programme for the Boeing KC-135E Stratotanker. This 1992 photograph illustrates the rather unusual sight of most of the remaining airframes at Davis-Monthan parked in very neat rows awaiting their disposal. Having yielded all the necessary parts for the KC-135E programme, as well as being of immense value in assisting other military or government funded projects, the airframes will eventually be tendered at auction and removed for scrapping.

Evergreen International, the Oregon-based US cargo carrier, operates a mixed fleet of Boeing 747s, 727s, DC-9s and DC-8 freighters on its worldwide route network. Two of its earliest DC-8 aircraft are still present at their Marana, Arizona, base almost ten years after being withdrawn from use. As the photograph shows, both aircraft have been extensively robbed for parts including engines, undercarriage assemblies and flight-deck windows, as well as most internal equipment. Both are series 52 models – N801EV, in the foreground was purchased new from the manufacturer by Air New Zealand in 1965 – and flew for the airline until 1977 when it was sold to Evergreen. A two-year lease to US charter operator Northeastern International followed in 1982, with the aircraft returning to Evergreen in February 1984, when it was withdrawn to Marana. The aircraft were still present in early 1994, though their continued existence would seem doubtful.

Boeing 707s and 720s were purchased on the commercial market by Boeing, on behalf of the air force, for the re-engining of KC-135A models with their more environmentally-friendly and economical Pratt & Whitney JT-3D fan engines. Once re-engined, the tankers were re-designated as KC-135E models for use by Air National Guard units, a number of which operate from civilian airports, thereby offering a significant benefit from having quieter powerplants in these ever more noise-conscious times. Along with the engines, a further ninety-seven line items were removed from each airframe to accommodate the modification of the KC-135A. As a lot of the airframes were bought by Boeing from aircraft brokers and not necessarily from their last operator, certain aircraft arrived wearing registrations not connected with the last airline that they flew for. One such example is Boeing 720-047B CX-BQG, which wears the distinctive colours of Air Malta with whom it flew as 9H-AAO. Disposed of by the Maltese carrier in January 1990 after being superseded by more modern equipment, the 720 was bought by the Boeing Military Company, Wichita, Kansas, on 31 August that year and flown to Davis-Monthan for parts recovery. Photographed in October 1992 looking fairly intact, the presence of a wooden trestle under the front fuselage would suggest the nose undercarriage is due to be removed soon.

More DC-8s at Marana, this time former Philippine Airlines examples, photographed in September 1981 not long after being withdrawn from service. RP-C837 was one of a number of ex-Philippine Airlines DC-8s retired to Marana in late 1980 for sale following replacement in the fleet by newer types. The series 51 model reverted to its original US markings of N808E, originally allocated when delivered new to Delta Airlines in 1962, following its sale to aircraft broker FBA Corporation in May 1983. The DC-8 went on to fly briefly for Aeromexico until it was again withdrawn from service at Mexico City and sold to a local aircraft parts company, where it was finally broken up for scrap in the early 1990s.

The Boeing 707s at Davis-Monthan have long outflown their commercial life cycles, or were near to it when acquired by the military, even though they were still able to yield millions of dollars worth of parts and equipment to the air force. The purchase price for the airframes was between $850,000 and $950,000 depending on condition, and the AMARC were able to recover over four million dollars in assets from every aircraft. Each aircraft to arrive at AMARC was marked for a specific tail number KC-135A at the modification depot at Wichita, Kansas, and all airframes had identical parts removed to provide for the modification. The 100 series 707 and 720 aircraft are more compatible with the KC-135A model, so more parts were removed from these airframes, although the larger series 300 aircraft were held to provide spares for the air force's E-3 and C-18 ARIA Project airframes (the latter themselves ex-American Airlines B.707-323Cs). Still adorned with the full markings of its previous operator, Boeing 707-338C, N342A last flew commercially as 60-SBM until sold to the military via Omega Air Inc. in May 1985, taking up its American marks on arrival at Davis-Monthan on April 2 1986. The undercarriage on the series 300 707s is interchangeable with that of E-3 Sentry and C-18 aircraft, and, as the photograph shows, all landing gear is removed and returned to a rework depot for overhaul, and put back into the active inventory.

Long before Mojave became a major holding site for some of the hundreds of superfluous airliners which presently blight the industry, it was home to the largest single concentration of Convair 880s anywhere in the world. American Jet Industries, well-known in the business as a major aircraft converter and spare parts supplier, bought up the remaining ex-TWA airframes which had been stored at the airline's Kansas City base following their withdrawal, which together with a few other examples were flown to Mojave in the late 1970s and parked up. This October 1982 view shows the impressive sight of nineteen Convair 880s and a solitary 990 stored on the northern side of the airfield long before it filled with more modern equipment, awaiting an uncertain future.

Former TWA Boeing 707-131B N86741 makes a sorry picture as it languishes in the yard of one of the scrap metal contractors which line the edge of the massive AMARC at Davis-Monthan AFB, Tucson. Retired from service in May 1982, the 707 was sold to the military and stripped of all useful parts by the air force to support their KC-135E conversion programme. Following the extensive parts reclamation within AMARC the airframes are made available to the General Services Administration for possible removal and transfer of parts to non-profit-making organisations such as community colleges, training facilities and the like. Once this commitment is fulfilled the airframes are turned over to the Defense Re-utilisation & Marketing Office (DRMO) for disposal. Sold at sealed bid auctions, the aircraft are towed off the AMARC site into the nearby contractors' yards, where the final scrapping process begins. The number 41-2160 on the nose refers to the particular sale this aircraft was entered in, and shows faded signs of having been an unsuccessful item in a previous auction. DPDO is the Defense Property Disposal Office (since retitled DRMO) and the code CZ029 an AMARC inventory number, given to each aircraft upon entering storage. Within AMARC the 707s are classed as C-137s, and, lacking any military serial, their allocated identity is taken from the aircraft's construction number; in this case 20057, as the last four digits attest.

Israeli charter airline Arkia sold two of its Viscounts in mid-1982 to the well-known Tucson-based Viscount specialist Go Transportation to act as a source of spares for other members of the fleet, joining an already considerable collection. Both aircraft, still wearing the registration and colour scheme of the Israeli operator, were parked in their new owner's compound and proceeded to donate parts to other more fortunate examples. Photographed in 1993, 4X-AVG, still wearing Arkia's old livery, was parked well away from the Go Transportation area, possibly impounded by the airport authorities judging by the very secure tie-downs attached to it. This particular Viscount has had an interesting career during the thirty-four years since it first flew, operating for Sudan Airways, British United, Aviaco, British Midland and Alidair before being sold to Israel in 1974.

Sadly, due to the overcapacity within the airline business at present as the industry slowly emerges from recession, many older aircraft which entered storage a few years ago will no longer return to flying duties. Ever more stringent noise regulations are now limiting the markets to which many of the used airliners can be sold, leaving their owners with few options for the airframes other than dismantling for parts. This ex-Continental Airlines 727 N88702 had had most of its useful parts removed when photographed at Mojave in October 1992, which at present rates would probably yield a better return for its owner in the aircraft spares market than a complete aircraft would be on the second-hand market.

The two-tone green fuselage stripe on this Boeing 707 gives a strong clue to the fact that it was once in the fleet of Dutch charter airline Transavia Holland. Its Dutch marks of PH-TVA were cancelled in May 1982 upon its sale to Guy-America Airways, taking up the registration N519GA. Used on services by the carrier between Georgetown, Guyana and New York, the airline later received the approval of the Civil Aeronautics Board (CAB) to use the name American Overseas Airways when it commenced worldwide charter services in June 1982, with a service from Boston to Athens. It was with these markings that the aircraft was bought back by Boeing in April 1983 on behalf of the air force, and ferried to Davis Monthan. It is shown in October 1992 after parts reclamation had been completed. Whilst in AMARC, all electrical wires are cut and throttle aisle stands on each aircraft are also removed in order to render it impossible for the airframes ever to fly again.

Looking like fallen tombstones, these are Boeing 707 and 720 tail fins in AMARC from some of the airlines whose time-expired airframes were obtained for the air force conversion programme of KC-135As into E models. The first three airlines to operate the famous 707 are represented in the photograph, with tail fins from Pan American and TWA and a large number of ex-American Airlines airframes shown in the background. Also in the picture is the fin of an ex-Jet 24 aircraft, a Miami-based charter carrier that ceased operations in the late 1980s.

Go Transportation, large-scale Viscount specialists, filed for bankruptcy in 1986, the result of a worsening financial situation brought about by a downturn in business. UK Viscount experts Jadepoint, owners of British Air Ferries, acquired the assets of the Go Group and attempted to revive the carrier with a complex recovery plan. This succeeded in the short-term, but sadly by 1992 the company was once again in difficulties, and its once proud fleet of Viscounts seemed a long way from ever flying again. Illustrated is series 797D N660RC, which was originally built for Capital Airlines but was never delivered; it eventually joined the fleet of the Canadian Department of Transport as CF-DTA in 1958, where it operated until its sale to Go Transportation twenty-four years later. During the early 1990s the assets of the company began to be disposed of, which naturally included the many airframes collected over the years. Sadly, a lot of the once-immaculate Viscounts were to make their final journey by road to the yard of National Aircraft, one of the nearby scrap metal contractors, where they were unceremoniously cut up. This included N660RC, which by October 1993 consisted of no more than a front fuselage section.

A total of 186 Boeing 707 and 720s were delivered to Davis-Monthan AFB between September 1981 and November 1990. By October 1993 this number had fallen to 102 almost complete airframes, two aircraft which had been mostly dismantled and the scrapped wreckage of another, with the yard of South West Alloys, one of the nearby scrap dealers, containing the better part of nine more examples slowly being taken apart. The lines of former civilian airliners contain examples from some less well-known carriers including this 707 N7158Z, last operated by Kenyan charter airline African Express Airways. African Express Airways flew the aircraft for a little more than two years as 5Y-BFF in this simple colour scheme, based on that of a previous owner, before selling it to Omega Air in July 1990, from whom it was acquired by the military.

Whilst most of the 707s at Davis-Monthan remain fairly intact during their time in storage, with only parts and components being removed, certain airframes are made available to the FAA for their use in simulations of the effects of an airborne explosion in a commercial airliner. Much of the testing is carried out by the Materials Laboratory of the Aeronautical Systems Division (ASD) to help improve the design of future commercial transports, with every airframe utilised to the maximum prior to being turned over to the DRMO for sale as scrap. N885PA, the former Pan American "Clipper Northern Light", has been extensively used in the FAA/ASD programme, and as the photograph shows is certainly past its best. Following some tests, certain sections of the fuselage have been cut out and removed so that a closer inspection can be made of the damaged area.

Some of the longer term inmates at Marana have donated certain airframe parts during their time in storage. Boeing 707-138B N792FA, inactive at Marana since mid-1983, has signs of a previous operator showing through the faded paintwork on the cabin roof. This aged airliner, the sixteenth example to roll off the production line, in August 1959, was seized by the IRS a few years ago in lieu of unpaid debts and has been sold for spares recovery. Minus its engines, the front cabin door assembly and overwing emergency escape hatches have been neatly removed, presumably for use on another aircraft.

Continental Airlines, one of the more fortunate American carriers, have survived the recession thanks to some adroit reorganisation and the help of a substantial cash infusion from a number of investors. A substantial number of their fleet were also financed on a "sale and lease back" arrangement which allowed the airline to capitalise on the value of its fleet and then leave the business of disposal of unwanted airframes to the lessor, who exchanged them for new types. The downturn in the financial fortunes of the travel business in the last ten years – coupled with the flood of newer, more economical types onto the market – has forced the value of older "first generation" jets down on the second-hand market. In order to realise their true worth, owners of such types now find it more profitable to dismantle the aircraft and sell the parts to other operators. Two examples of this are seen here at Mojave in October 1992, as ex-Continental Boeing 727s being "parted out". Most of the useful parts have been removed, including all instruments, windows, doors, wing control surfaces, flap tracks and undercarriage doors, together with areas around the emergency exits which have been cut out. N18480, which first flew in August 1964, was retired to Mojave in July 1991.

The US Air Force's KC-135E conversion programme involved the replacing of original, noisy and thirsty J-57 turbojets with JT-3D fan engines acquired from former commercial 707 aircraft. Performance of the re-engined tanker is significantly improved, as well as being more economical, with the new power-plant also requiring less maintenance than its predecessor. The complete programme involved slightly more than simply swapping engines, and included various modifications to the airframe and electrical systems to accommodate the changes. Its hydraulic system drained, evidenced by the drooping flaps, this former Guyana Airways 707 has donated its valuable engines to the KC-135E programme, and will also yield thousands of dollars worth of other parts in the form of internal equipment which can be used on other air force aircraft. Each Boeing 707 purchased for the programme has given a staggering $4 million worth of parts, representing a 500 per cent return on their purchase price. This particular 707, N1181Z, an original Pan American aircraft, arrived at Davis Monthan on 20 February 1986, and was still present eight years later.

Hamilton Aviation of Tucson are renowned for their expertise with the Convair series of airliners, and have built up a huge stock of spares through the purchasing and dismantling of airframes which have been withdrawn from service. Other types are also catered for, in particular the Japanese NAMC YS-11, itself a competitor for the Allison-powered Convair 580. As a consequence, at least one example is held by Hamilton for spares recovery as this October 1991 photograph illustrates. N903TC still wears its colourful Simmons Airlines livery (with whom it last flew) having been withdrawn from service in October 1986, following which it was sold to Short Bros (USA) as part of a purchase deal on new aircraft. The YS-11 had arrived at Tucson by March the following year and, although it has had two different owners since then, it has never flown again.

Dubbed "the world's biggest glider" by staff at the Evergreen Air Center, this Boeing 747 is situated in a remote location on the far side of the huge Marana base of Evergreen International, and is firmly grounded thanks to its being bolted to three huge steel trestles. One of nine 747s on the fleet list of Evergreen, N476EV was built for the type's launch customer Pan American in 1970 as N751PA, making its first flight on 6 April that year. Eighteen years of faithful service followed before it was sold, joining the ranks of competitor TWA. This move was short-lived, however, as in January 1989 it was purchased by Evergreen International, where it spent a period of lease to holiday companies Orion Air and Air Europe flying charter services from the UK. The 747 was returned to Evergreen in December of that year, but by mid-1991 it had been retired from service and parked up at Marana.

The Boeing 747's amazing transformation from "jumbo jet" to "grounded glider" took place during early 1992. After undergoing extensive parts reclamation by company engineers, the airframe was towed across the airfield on only three of its five original undercarriage assemblies and with all engines removed. The airframe was positioned above the series of heavy-duty steel trestles, which are sunk deep into the ground, and the remaining landing gear was removed, resulting in this rather bizarre sight. Even though it has been robbed of a substantial amount of airframe parts the 747 is still potentially airworthy and could be returned to flying status should the need ever arise, this being part of the reason it has been preserved in this manner. At the time of this photograph (October 1992) the airframe had been used on occasion as a valuable training aid by the Federal Law Enforcement Agency for anti-terrorism training on a modern widebody airliner.

Probably the first of many . . . the sad sight of the first of a number of series 100 Boeing 747s which are due to be scrapped in the next few years. The Boeing 747 is now well over twenty-five-years-old, the first example having rolled out of Boeing's Everett plant at Paine Field, Washington, on 30 September 1968 in front of a jubilant workforce, astounding the world's press and, more importantly, representatives from the world's leading airlines who planned to re-equip their fleets with this amazing aeroplane. The 747's design dates back to a 1964 requirement by the US Air Force for a heavy logistic transport aircraft, known by the programme name CX-HLS (Cargo Experimental-Heavy Logistics System). As is now well-known, Lockheed won the contract with what would become the C-5A Galaxy, but Boeing now had its sights set on the commercial world. N749PA, the thirtieth 747 to come off the production line, spent its whole life flying for Pan American, delivered as "Clipper Intrepid" on 10 April 1970. Originally owned by the airline, N749PA was re-financed through banks and commercial industrial creditors following Pan Am's need to improve its financial situation of the late 1980s, but continued to fly for the airline. Following the collapse of Pan American in December 1991 the 747 was immediately returned to its owner, Citibank, and withdrawn to Marana for storage. In June the following year aircraft parts company Amtec Jet bought the airframe and commenced the task of dismantling the aircraft at Marana, alongside sistership N750PA which had followed an identical route.

Still wearing its owner Evergreen International's full colours, N476EV shows the level of parts recovery to which it has succumbed, with all wing devices and engines removed, together with numerous cabin windows. The airline holds sufficient spares in its inventory to enable this 747 to be fully rebuilt, although with the current surplus of similar airframes this option will probably not be taken up.

This photograph illustrates the fate of all the Boeing 707s which were held in AMARC as part of Boeing's KC-135E programme for the USAF. Once parts reclamation has been completed on site, and the aircraft has been earmarked for disposal by the DRMO (Defense Re-utilisation & Marketing Office), eligible buyers are invited to inspect the airframes and submit a tender in a sealed bid auction. Sold as lot #15 in auction number 41-2160, the ex-American Airlines 707 shows signs of a previous auction entry as lot #54. This may be because of an unsuccessful bid, though it is more likely that the contractor could not remove the airframe from the facility in the given timeframe and consequently forfeited his contract. N7552A, a series 123B aircraft, spent its entire career flying for the airline since its delivery in May 1965, and was retired to AMARC in February 1983. Ironically, it was American Airlines who played a significant part in the introduction of the Pratt & Whitney JT3D turbofan engine when their first "B" model (as this sub-type was classed) was delivered in June 1960, an engine that would prove so popular with the air force some twenty years later.

Many of the ex-civilian Boeing 707s parked in the AMARC at Davis-Monthan AFB wear the colours and markings of long gone carriers, giving the area a strange connection with the past. GIA stood for Global International Airways, a US passenger charter airline which flew to destinations in the Caribbean, the United States and many European cities during the late 1970s and early 1980s. Sadly, the name Global was to disappear in December 1983 when financial difficulties forced the suspension of its operators certificate by the FAA. The fleet was made up of former Pan Am and American Airlines airframes which were either returned to their lessor or sold to other charter carriers, with the majority eventually ending up at Davis Monthan. This example, N8417, was originally delivered to American Airlines and once flew for British Caledonian as G-AYZZ. It had joined the Global fleet by April 1981, originally on lease, and was eventually purchased by the airline two years later. Repossessed after Global's demise, it arrived at AMARC in November 1985.

Ethiopian Airlines' distinctive colour scheme has been seen on both Boeing 707s and 720s over the years, receiving its first examples in 1968. Nowadays the carrier's name is seen around the world on more modern equipment, although still from the Boeing stable, with a mixed fleet of 757 and 767s now flying the long-haul services, and 727 and 737s operating the shorter routes. Boeing 720-024B N550DS was bought from Continental Airlines in 1974 and flown as ET-AFK with Ethiopian Airlines; following retirement fourteen years later it was flown initially to Marana for storage before being purchased by Boeing for use in the ANG tanker re-engining programme. Photographed in October 1992 at Davis-Monthan, the major components for the conversion have been removed, though the aircraft remains otherwise intact.

Mojave Airport in California has long been known for the many unusual aircraft likely to be seen there at any one time, both operational and non-flying. This very early Boeing 707-131 N198CA owned by Charlotte Aviation, photographed in October 1992 , was the twenty-second model produced, and has been parked at Mojave since November 1978 in the company of N194CA. During that time both have remained virtually intact while those around them have been dismantled and scrapped. Both aircraft have led quite interesting lives, having been delivered originally to Trans World Airlines in April 1959. N198CA wore the registration N734TW when in service with TWA and was withdrawn and replaced by a more modern variant of the successful 707 in 1971. Its sale to a Philippine operator, Air Manila, via a company called the Cranbourne Corporation in 1971 fell through, and the Boeing was returned and later took up the registration 4X-AGT upon sale to Israeli Aircraft Industries in January 1975, from where it was transferred to the Heyl Ha'avir (Israeli Air Force) four months later and re-registered 4X-JYI. In IDF/AF service the 707 has replaced ageing Boeing C-97 Stratocruisers, and is flown as more than just a transport with some airframes converted for the tanker role as well as for various ELINT and SIGINT missions. 4X-JYI wore the air force's simple but rather distinctive blue and white livery, presumably in an attempt to resemble a civilian airliner when on overseas flights, with the stylised blackbird and blue globe of 120 Squadron applied to the tail fin and repeated on the nose. This aircraft probably flew in the transport and VIP role from the Unit's Ben-Gurion International Airport base, and was disposed of in November 1978 after being superseded by newer, longer range 707 models. Charlotte Aviation was an aircraft broker and trader outfit specialising in the refurbishment and sale of second-hand 707s, and after retirement from the IDF/AF, the Israeli registration was cancelled and the marks N198CA applied. Flown to Mojave soon afterwards, the 707 was stored, with its ex-IDF/AF stablemate N194CA, in a separate fenced compound on Mojave's north side, its ex-military insignia painted out.

Of all the Viscounts belonging to Go Transportation stored at Tucson in the mid-1980s, the only example ever to wear any airline markings was this Mexican XA-MOS. In a smart red and blue scheme, the series 745D was last flown by Mexican charter operator Aerolineas Republica, from whom it had recently returned when photographed at Tucson in October 1982. The former Capital Airlines Viscount was acquired by the Go Group in March 1977 and registered as N220RC; this followed a short ownership by the Loving Insurance Agency as N500TL and, before that, it was the personal transport of singer Ray Charles, who flew it as N923RC. By the time of the photograph the aircraft had donated its number four engine to another Viscount, though sadly the next ten years saw XA-MOS giving up more and more parts to other more fortunate fleet members, and by October 1993 this immaculate airliner had been cut up for scrap in one of Tucson's many aircraft salvage yards.

The perimeter of the huge AMARC at Davis-Monthan AFB, Tucson, is ringed with numerous surplus aircraft salvage yards, to which former inmates of the storage facility are periodically delivered. As a consequence, with the exception of the ex-airline 707s, the contractors normally deal in former military airframes. One yard, however, does have the distinction of containing some rather interesting civilian aircraft, which look quite out of place. Within the compound of DMI Inc. are four examples of the Australian-designed and built GAF Nomad, which last flew for Coral Air. N420NE and N421NE were part of a fleet of Nomads, bought from Hughes Aviation Services in 1980, which were repossessed after the demise of Coral Air in mid-1982 and initially stored at Las Vegas-McCarran. In the late 1980s four of the aircraft were purchased by Don Howell of Tucson, who had them transported to DMI with plans to restore and sell them. However, in October 1993 all four examples were still derelict in DMI's yard.

Giants at rest. Resembling an aviation equivalent of an elephants' graveyard, the north end of the main ramp at Pinal Air Park, Marana, was the final resting place for at least two Boeing 747s during 1992, and in October two more examples were parked ominously close by. The type is now well into its third decade of operation and, due to the almost unmatchable qualities the 747 possesses, the type has been flown to its maximum by most of the western world's major long-haul carriers.

Consequently, a lot of the early production airframes are fast approaching the end of their flying career as they reach their design fatigue limits. Many former Pan American aircraft (the type's launch customer) have now flown extensively since 1970 in some cases for a number of different carriers following Pan Am's demise, and are now the first examples to face the scrap man's blow torch. Evergreen International was one operator to buy early ex-Pan Am 747s, with seven of their fleet

of sixteen aircraft coming from the airline. This October 1992 picture shows two early series 121 models in an advanced state of parts reclamation at Marana, following their retirement from service with Evergreen. N483EV and N484EV sit on only two of their four original main gear assemblies, thanks to the airframes being considerably lightened by the removal of engines and internal fittings.

A close-up of N484EV, stripped clean of all paint, shows obvious traces of its original fleet name with Pan American, "Clipper Arctic", which was applied to the aircraft when delivered new in June 1971 as N657PA. Bought by Evergreen in May 1988 it was immediately leased back to Pan Am for the next three years, and eventually joined its owner's fleet in April 1991 when it took up its new registration. Following its return to Evergreen the 747 was flown to Marana for storage, pending a decision on its future. The aircraft was robbed of all the parts which could possibly be useful on the remainder of the Evergreen 747 fleet, and in order for the airframe to maintain a safe centre of gravity, huge concrete-filled cases were attached by heavy chains to its engine pylons. The aircraft is obviously of considerable value to Evergreen, as by summer 1993 both this 747 and N483EV had been moved across the airfield and stored atop huge steel trestles with the similarly configured "glider" N476EV.

The very sad-looking remains of this airliner, photographed in the yard of South West Alloys, Tucson, in October 1992, was once a proud member of TWA's large Boeing 707 fleet, where it flew as fleet number 5711. Delivered new to the airline as N15711 on 27 June 1968, the aircraft helped to consolidate TWA's worldwide reputation as one of America's leading operators. In May 1979 N15711 was acquired on lease by Guinness Peat Aviation, who immediately sub-leased it to Pakistan International Airlines. Returned to GPA within ten months, a two-year period of lease then commenced with Haitian operator Air Haiti, for which it was repainted in the carrier's red and blue colours, which it still wears in the photograph. Returned to TWA in the summer of 1982, the 707 was initially stored at Kansas City before it was obtained by Boeing Military Airplane Company, via agent Citicorp, and flown to Davis-Monthan AFB for use in the KC-135E re-engining programme. The very neat way in which the aircraft's nose section has been taken off, and the re-sealing of the open fuselage would suggest it had been removed whilst still in AMARC, possibly taken for future use as a procedure training simulator.

The legend on the nose of this former Ports of Call Boeing 707 spells its eventual fate, "disposal" meaning that the aircraft has been released by the DRMO for sale as scrap. The attractive colours of the Denver-based operator are slightly corrupted by the aircraft having been fitted with the cabin door from an American Airlines example. Whilst in AMARC the 707 airframes are made available to the FAA, NASA and the FBI for experimental purposes and use in various exercises, which could account for the door swapping on certain examples. N701PC was flown into Davis-Monthan on 4 August 1987, and as a series 123 model has undergone extensive parts recovery for both the KC-135Es of the Air National Guard and the air force's KC-135A model, with which the 100 series Boeing 707 has much in common.

The ubiquitous DC-3 has long confounded aircraft designers, who have struggled to design and market a suitable successor for this ageing, yet irreplaceable type. Many have tried, yet few have been able to provide an alternative which can offer such flexibility and versatility as the DC-3 at a cost many smaller operators can afford. An obvious alternative was to offer an improved version of the primary design, with the original radial engines replaced by more powerful turbine engines which would improve performance, reliability and the cruise speed of the DC-3. Early experiments in re-engining the DC-3 proved mildly successful, but it was not until the mid-1960s that the idea was considered seriously. Mr "Jack" Conroy, the well-known founder of Aero Spacelines whose inventive mind was responsible for the amazing Guppy and Super Guppy series of aircraft, obtained a former United Airlines Viscount 745D

from which he transplanted two Rolls-Royce engines into the DC-3's re-designed nacelles in the May 1969 conversion. Under the new name of "Turbo 3" Conroy hoped to market the new type to the airline world, a type that could boast a cruise speed in excess of 210 mph and a range of 2,250 miles, though sadly it was greeted by a very lukewarm response from prospective customers. Undaunted, Conroy reorganised under the new name of Specialised Aircraft Corporation, taking his original Dart-powered "Turbo 3" and further converting it to an even more radical design. This would produce the strange-looking "Tri Turbo 3" which had new Pratt & Whitney (Canada) PT6A-45 turboprops fitted on each wing position, together with a third faired into the nose of the DC-3. First taking flight on 2 November 1977, the "Tri Turbo 3" was marketed extensively once more, though again to no avail. Registered N23SA, the aircraft was

eventually painted in the high visibility colours of Polair Inc. in August 1979, and spent some time flying operations in the Arctic fitted with skis on the undercarriage. By October 1991 the aircraft had been retired and flown to Mojave where its three engines were removed in anticipation of the scrap man, as the photograph shows. Fortunately, a reprieve for N23SA was given, and in an ironic twist it was bought by Basler Turbo Conversions, who now successfully convert and sell the BT-67, a twin turboprop, stretched conversion of the DC-3 powered by two PT6A-67 engines. The DC-3 was dismantled and trucked out of Mojave to Basler's Oshkosh, Wisconsin, base on 17 October 1993, and after a lifetime of bad luck this DC-3 will now be converted to turboprop power for the third time.

Once parts removal by the air force within AMARC is completed, the remaining 707 hulks are available for disposal at auction by the DRMO. Due to the lack of available space within the facility, it is imperative that the aircraft to be removed are taken away by the successful bidder as soon as possible and that the contractor does not forfeit his contract with AMARC, which specifies a determined time limit for aircraft removal. As a good number of 707s have had their undercarriage removed

prior to being disposed of, particularly the 300 series models whose assemblies are compatible with C-18 and E-3 aircraft, the contractor must reattach nose and main gear units to these airframes in order to move them out of the AMARC. Consequently, a supply of spare under-carriage sets is held by the contractors, those which have been removed from earlier 707s disposed of in a more complete state. One of the major recipients of the ex-civilian airliners to come out of

Davis-Monthan has been National Aircraft, who have the airframes removed to the yard of South West Alloys situated on Kolb Road, Tucson, across the street from the AMARC site. Here, six sets of Boeing 707 main gear bogies are pictured in the yard awaiting their next customer with the remains of a former Ethiopian Airlines example in the background.

The introduction of the Boeing 707 into commercial service in 1958 heralded a new dawn in air travel. Grasping the limelight from the unfortunate Comet, the 707 was to become the international flagship for many airlines around the world for the next decade, until the introduction of the first generation widebodies in the 1970s. The classic type was built in eight different variants, and

was to wear the colours of most western national carriers as they converted to jet operations from the DC-6s, DC-7s and Super Constellations which had pioneered their post-war intercontinental routes. The 707s in AMARC show a fair selection of schemes from a number of the carriers to have flown the type, some of which are no longer is existence. This October 1992 picture shows examples from Air Malta,

TWA, Jet 24 and Israeli flag carrier El Al. TF-AYG flew for its whole career as 4X-ATT, forging links between Tel Aviv and most major western capitals for state airline El Al. Retired and stored at Brussels following its sale, the aircraft moved on to Davis-Monthan to donate parts to the USAF in October 1989.

The famous Pan American name was last seen on a Boeing 707 in regular service with the airline in 1981, when N881PA flew between New Orleans and Philadelphia on 26 January of that year. The routes which had previously been the domain of the type had been given over to 747s, TriStars and DC-10s by this time, which were operating in significant numbers by then. However, in a nostalgic move by the airline an example of the type was repainted in full colours to re-enact the type's inaugural trans-Atlantic scheduled service, between New York and Paris-Le Bourget, on 26 October 1983, the twenty-fifth anniversary of the event. N880PA had been withdrawn from service with Pan Am on 26 January 1981 with a total airframe time of 35,319 hours, and following a short period of storage was leased to Guyana Airways for use on their Georgetown-New York route. Returned from lease in December 1982, the aircraft was repainted in suitable period markings and given the name "Clipper America" after the original series 121 which began the airline's jet services. In a very stylish move, Pan American managed to locate a lot of the passengers and crew from the original event to help celebrate this rather significant milestone in commercial aviation. Sadly, however, the aircraft used on the original flight, series 121 N711PA, was disposed of by Pan Am in the mid-1970s, and was eventually broken up in Taipei in 1984. Meanwhile, N880PA's moment of glory was short-lived as the aircraft was retired once again upon its return from this celebration, bought by the Boeing Military Airplane Company in February 1984, and flown to Davis-Monthan on the twenty-eighth of that month to join the umpteen similar type in the desert.

Part of the aircraft "boneyard" situated on the north side of Mojave Airport, photographed in October 1990. This fenced-off compound contains many airframes both complete and in pieces which have languished at this quiet Californian airfield for a number of years. A number of the aircraft parts come from types which have been involved in minor accidents, written off by the airline as uneconomical to repair, though they were still reasonably intact at the time of their disposal. The contents of the compound are owned by a local aircraft spares company, which has developed a lucrative business in supplying aircraft, both whole and in parts, to the movie industry. This view shows some of the original "desert airliners" that had been here long before the arrival of the hordes of recessionary retirements which have filled the acres of sun-baked hardstands at Mojave in recent times. Two retired DC-8 series 21 aircraft sit alongside the remains of an ex-Allegheny BAC 1-11 and a Boeing 727 last flown by National Airlines, both of which suffered accidents in 1978.

The successful family of Convair airliners has earned a reputation as an economical and easy-to-maintain type, well-suited to short and medium range routes both in passenger and freight configuration. In recent years, two well-known companies in North America with long associations with the type have been developing a Convair "successor", by improving the original airframe into a more cost-effective aircraft. The programme involves lengthening the fuselage and in certain cases re-engining with Allison turboprops,

and significantly updating the flight-deck with state of the art instrumentation. Kelowna Flightcraft in Canada have now successfully converted one airframe into the Convair 5800, stretching and re-engining an ex-US Navy C-131F, which it is hoped will be the first of many. Similarly, Tucson-based Hamilton Aviation has collected a stockpile of former Wright Airlines and Kittyhawk Convair 600 and 640 aircraft over recent years, with a plan to utilise the airframes in a similar programme for a "Super 580"

conversion. Examples of these are seen withdrawn from use at Tucson in October 1991, with engines and outer wing panels removed, but otherwise complete, awaiting a decision on their future. The project would appear to have been shelved however, as by early 1993 all of the retired aircraft which could have been used in the conversions had been broken up by a local scrap dealer and removed, throwing doubt on a rival for the Canadian design.

The plain and simple scheme on this Boeing 707 identifies it as last being flown by Air India, who flew the type on its international route network until the mid-1980s. At one time the Indian national carrier's fleet consisted entirely of 707s,

but the type was eventually superseded and replaced by 747s and examples of the European Airbus family. N8880A was delivered new to the airline as VT-DPM in May 1964, and was eventually retired almost twenty-four years later. A short

lease to an operator called Atlanta Icelandic followed, registered TF-IUE, before it received its American registration. It arrived at Davis-Monthan in October 1989.

The need to store the many hundreds of redundant airliners in the deserts of southern California and Arizona has become big business for some, with the three or four main sites competing strongly against one another for numerous fleet retirement deals which have swelled the numbers of aircraft parked up in recent years. This situation can lead to some rather unusual sights, such as lines of brand-new airliners flown directly from the production line to the desert storage site, or complete airline fleets parked up side by side following the untimely demise of their owners. This surreal view, at Marana in October 1993, shows that even the mighty 747 is not invincible. These are the last remains of two former Pan American aircraft which had been retired in 1991 following the collapse of the airline. Sold by Citicorp, their registered owner, the 747s, N749PA and N750PA, became the property of an aircraft parts company Amtec Jet Inc. who proceeded to dismantle them during late 1992. A chance comment by an employee, who suggested that the distinctive 747 "humps" could possible be of use to a movie company, secured some form of preservation for these remains, which were moved into the centre of the airfield and were awaiting sale when photographed.

Once part of the United Airlines fleet, as the registration N8015U denotes, this DC-8 was built in 1959 as a series 11 model, one of twenty-eight examples of this original variant which were produced for use on the US domestic routes of Delta and United Airlines. Bearing the name "Mainliner Capt. W. E. Rhoades" the DC-8 helped to establish United's jet aircraft services in the early 1960s replacing the now outmoded piston-powered DC-6s and '7s in their fleet. Re-engined with Pratt & Whitney JT4A turbojets in 1966, which converted the aircraft into a more powerful series 21, N8015U operated with the carrier until its retirement in June 1977. Purchased soon afterwards by Casino Royal Inc., the DC-8 was transferred after three months to an outfit called the Sundance Travel Club. Operated for a while as the transport for a number of different travel clubs, the aircraft was eventually withdrawn from use in 1980, and initially stored at Burbank Airport. Bought by American Jet Industries (later renamed Gulfstream American Corp.) for use as a spare parts source, the DC-8 was flown to Mojave on 23 March 1982 and joined a number of other airframes also owned by AJI, including another DC-8 which was last flown by Eastern Airlines, all in various states of disrepair. After being robbed of all useful parts, the airframe, minus its engines, became the property of a company specialising in the supply of aircraft parts for use in film and TV movies, helped, no doubt, by Mojave's proximity to the film industry capital Hollywood. Photographed in October 1990, traces of previous operators are visible on the cabin roof, with the titles of Fiesta Air and Florida Air Travel discernible through the fading white paint.

Within two years, both DC-8s which had been derelict at Mojave were dismantled and transported the short distance to the premises of their new owner, and preparations were begun for their future "starring roles". This former Eastern Airlines example, N8604, was the forty-first of the type to be built, making its first flight from Long Beach in February 1960 prior to delivery to its new owner later that month. The DC-8 operated for the airline from their Miami base for the next thirteen years, before its sale to a leasing company in September 1973 from where it was supplied to Air Haiti soon after for a short period of lease. Purchased by American Jet Industries in 1977, the aircraft ended up at Mojave and remained parked for the next fifteen years before experiencing a similar fate to N8015U. Photographed in the yard of its new owners, the front fuselage of N8604, still adorned in the distinctive two-tone blue livery of Eastern Airlines, sits behind that of N8015U in October 1992, with a DC-3 parked beyond. The black marks painted on the fuselage side to simulate fire damage bear witness to the fact that the DC-8 has already seen service in its new role.

The Hollywood movie industry has benefited prodigiously from the abundant supply of derelict airframes which have been withdrawn from use at Mojave over the years, with aircraft parts both large and small making the short journey from the airfield to the yard of the "props" supplier for use in numerous TV programmes and films. This Boeing 727 front fuselage shows the scars from its use as the centre piece of the recent Dustin Hoffman hit movie *Accidental Hero*, where an airliner crash scene had to be recreated for the cameras. Originally built for the Peruvian airline Faucett in 1968, where it flew as OB-R-902, the 727 was bought back by the manufacturer in May 1985 for use in conjunction with engine-maker General Electric's "Ultra High By-Pass/Un-ducted Fan" engine which was fitted to one of the airliner's side engine positions. The aircraft was returned to Boeing following completion of the engine test programme in early 1987 and flown to Mojave for storage. Following a repaint into the fictitious markings of "Midwestern Airlines", the 727 was sectioned at Mojave and removed by flatbed truck, and received the necessary "make-up" and suitable airframe damage to simulate the crash. Two 727 front fuselages, together with the remainder of this aircraft were actually prepared for the film, and following the completion of shooting at the crash site the aircraft sections were returned to the yard, where they will probably be repainted and made available for future use in another production.

Located at the junction of Highways 14 and 58 some ninety miles north of Los Angeles, the town of Mojave is well-known as the home of the civilian flight test centre, as well as being the birthplace of the famous Burt Rutan-designed "Voyager" airplane, which made history as the first aircraft to fly around the world without refuelling. The vast airport has seen many unusual aircraft come and go over the years, though there are still a number of interesting types to be seen. Two former US Air Force Douglas C-133 Cargomasters continue to sit quietly in the baking sun following their ferry flight from Davis-Monthan Air Force base in 1975. From a total of six examples sold onto the US civil register, the two aircraft were part of a fleet of four Cargomasters originally destined for relief operations by the non-commercial Foundation for Airborne Relief (FAR) and were given the civilian registrations N136AR and N201AR. Intended for use alongside other ex-military types such as C-97 Stratofreighters in various famine and disaster regions around the world, the plan did not come to fruition. The other two examples, intended for this work were flown to Tucson and Long Beach where they were stored, the latter example eventually being scrapped in the late 1970s. Of the remaining two C-133s, one was purchased by Alaskan freight operator Northern Air Cargo in 1973 and was later sold to the aptly named Cargomaster Corporation, who had set up operations with the other C-133 in 1977. The three surviving Cargomasters, at Tucson and Mojave, were then purchased by Cargomaster Corp. to provide a welcome supply of spares for the Alaskan airplanes, of which only one is now operational. The two Mojave Cargomasters are still fully intact, with engines still attached. This October 1991 picture shows N201AR, the former 56-2001, with all military markings scrubbed out and without its rudder, which was removed following slight damage caused by strong winds.

Another TV star at Mojave, this rather unauthentic looking DC-8 series 33F was painted to represent a military transport type for a TV recruitment commercial on behalf of the Air National Guard. It also starred in a cameo role in the spoof film *Hot Shots* which was also filmed at Mojave. Built in 1961 for Pan American, the order was not taken up and the DC-8 was eventually delivered to Panair of Brazil as PP-SDS. Transferred to VARIG in 1965, it was later bought by the Brazilian government and leased back to VARIG before being finally withdrawn from use and stored at Porto Alegre in October 1975. Bought by American Jet Industries three years later, it was given the registration N59AJ, eventually ending up at Mojave in 1981 where a further period of storage ensued. It received its military "uniform" in early 1987, along with another example, in readiness for its TV appearance, and was returned to storage again once filming was complete. This October 1990 photograph shows the DC-8 being readied for flight once again following its sale to Columbian freight operator LAC Columbia, with whom it was registered in August 1990 as HP-1166TCA. It finally departed Mojave on 2 March 1991 for Phoenix, and flew onwards to Miami where it was repainted in full LAC Columbia scheme pending entry into service.

The Convair 880s stored at Mojave since 1979 have also made occasional starring roles in a number of films. One example was repainted in the colours of fictional airline "Pan West" and given the false registration N375 for its part in the TV film *Amazing Stories* in 1986, and another was purchased by Warner Bros in 1990 and dismantled and removed for use in a different project. This Convair 880 was one of the fifteen former TWA examples which were stored at Mojave in 1979 together with four unmarked aircraft, the property of American Jet Industries. A simple colour scheme change was given to the aircraft with its distinctive TWA cheatline overpainted in blue with a yellow trim, and matching tail markings. Some minor engineering work was needed to assist its scene in the recent Clint Eastwood movie *The Rookie*, which required the port wing tip to detach following a collision on take-off with a Hansa Jet, which was pursuing the stars of the film across an airport at night. The Convair was taxied down the runway and duly made contact with the executive jet, which was virtually destroyed in the scene, and the fabricated wing end on the 880 came away amid much sparks and pyrotechnics. Following its movie appearance the original wing section was reattached with the aid of two rather crude bracing struts, and the aircraft, now wearing an engine panel from a Delta example, was towed back to the compound of derelict airframes at Mojave's north side to await another job.

The paint may be faded, the windows broken and missing and all of the vital parts of the aircraft removed, but Convair 240 N51331, photographed in the desert compound of a Californian company in the business of supplying aircraft parts for use in movies, may still have a future as an "extra" in a film or TV programme. A very early model 240, delivered new from the San Diego production line in June 1948 to American Airlines as NC94238, it operated in these markings until 1958 when it was sold to an aircraft broker, and sub-leased to Continental Airlines for six months. Several other ownership changes over the following few years took the Convair to Japan, where it joined the fleet of TOA Airways for three years as JA5110. Returning to the US in April 1965 it took up the marks N51331 with the Miami Aviation Corporation, from whom it was sold to Cordova Airlines in June of that year. This operator merged with Alaska Airlines in early 1968 and the Convair received the colourful scheme of the new airline, which it would wear through a further series of leases to other carriers, until retirement in 1976. Stored at Long Beach, the Convair was in the company of a number of other retired and repossessed airliners when it was sold and broken up in the late 1970s. Its move to the Californian desert ensured its continued existence, alongside other similarly acquired airframes.

Rather unusually for a 707, this aircraft spent its entire career flying for one operator, having been delivered new to American Airlines in September 1966 as N7564A. At their peak in the early 1970s, American Airlines operated one of the largest collections of the type in the world, before the many widebody types such as the DC-10 were added to their fleet. This particular aircraft experienced a similar fate to so many others in the mid-1980s; withdrawn from use by the airline in June 1984 it joined nineteen other former American Airlines examples in storage at Davis-Monthan for use in the air force's KC-135E programme. Surprisingly, this former American Airlines 707 had its colour scheme removed prior to disposal from AMARC, and is seen in the yard of contractor South West Alloys surrounded by the accumulated debris generated by the final destruction of the aircraft. A stipulation of the sale agreement to the contractors is that no airframe parts are authorised for sale to prospective civilian customers, and all such trade should be processed via the DRMO. This 707 has donated its cabin door assembly as well as its side cargo door, and the pile of passenger seats alongside illustrates that many aircraft were retired and stored at AMARC in full commercial fit.

Because a lot of film and TV appearances by Mojave's derelict aircraft never require the full airframe, many uses can be made of one particular aircraft. An upper fuselage section from former United Airlines DC-8-21 N8015U is shown loaded onto a trailer at Mojave in October 1992, after the remainder of the aircraft had been removed. This section will probably be repainted in more fictitious markings for use in another film. The very uncertain future of a lot of the older "stage 2" airliners presently stored at Mojave, such as the DC-9s and Boeing 727s in the background of this picture, will eventually lead to some of their numbers following a similar path to the DC-8 here. By October 1993 a 737 and two 727s from Continental Airlines and a former Eastern Airlines DC-9 had already begun to be dismantled in this compound, as they have now become more valuable in pieces than as a whole.

The Dutch aircraft manufacturer Fokker was one of a number of companies that had foreseen a good market for a fast, modern, pressurised aircraft which would be needed by many operators as a DC-3/C-47 replacement in their fleet modernisation plans of the 1950s. The Rolls-Royce Dart-powered F27 design became a very popular choice with many operators, and was widely accepted in the competitive North American market, so much so that American manufacturer Fairchild secured an agreement for licensed production of the type, which ironically was to beat its Dutch equivalent into service in July 1958. Air Cortez were a US third level operator who flew a fleet of three Fairchild F-27s out of Ontario Airport, California, to a number of Mexican destinations, as well as Grand Canyon and Las Vegas. Sadly, the company failed in the early 1980s and its aircraft were returned to their lessors. F-27 N726US was originally operated as N2779R by Pacific Airlines in 1964, who merged with Air West in 1968 to form the new carrier Hughes Air West. Bought by US Aircraft Sales in the early 1980s and leased to Air Cortez, it took up its present registration, and is shown at Las Vegas-McCarran in October 1982, being dismantled after its retirement, following return from Air Cortez.

A one-time Mojave resident, this BAC 1-11 sits in the premises of a Californian aircraft spares company, part of a huge collection of airframes both whole and sectioned which are held for possible future use in a TV or film production. Still wearing the fleet number and colour scheme of its last owners Allegheny Airlines, N1550 was damaged beyond repair in July 1978 when it overran the runway whilst landing at Rochester Airport, New York. Deemed as uneconomical to repair by the airline, the aircraft was cut up and removed, with major sections of the fuselage transported to the "boneyard" compound at Mojave Airport. All internal fittings were removed from the 1-11 prior to its move to Mojave, though sufficient spare instruments and internal equipment are held by its present owner to recreate an authentic airliner mock-up.

Second only to UK operator British Air Ferries, the Go Transportation Group were one of the last major operators of the Vickers Viscount, accumulating over thirty examples of the type during their peak in the 1980s. Sadly, the company's fortunes waned towards the end of the decade, and its once proud fleet lapsed into disrepair as they flew less frequently towards the end. Photographed in October 1991, the Go Transportation ramp at Tucson contained most of the remaining airframes, withdrawn from use and parked tightly together awaiting an all-too-certain fate. Years of storage in the open air at Tucson have taken their toll on Viscount 827 N480RC-F, which was delivered new to Brazilian carrier VASP as PP-SRC in 1958. Sold in the mid-1970s to Uruguayan national airline PLUNA, it became the property of the Go Group in March 1982 and was delivered to Tucson wearing its American registration. Since then the aircraft has not flown, but has slowly contributed many parts to other more fortunate fleet members.

Pinal Air Park, Marana, has seen significant numbers of retired airliners arrive for storage in the recent past, and as a consequence some of the older long-term residents have had to be moved to accommodate the latest arrivals. Boeing 707-138B N792FA was flown into Marana in mid-1983 from Burbank Airport, where it had been in storage since 1979 after its last commercial flight for Indonesian airline Bouraq. This very early model 707 was built for Qantas in 1959, who operated it as VH-EBF "City of Adelaide". It continued in these marks for the next nine years before the aircraft was purchased by brokers F. B. Ayer & Associates, where it was given the registration N792SA. A series of short leases to numerous different operators commenced in March 1968 which saw the aircraft flying with Standard Airways, Air Commerz of Germany (who operated it as D-ADAQ) and Turkish national airline THY, where it received the markings TC-JBP. Returned from this lease, it took up the new registration of N792FA, and was repainted in Indonesian carrier Bouraq's scheme for its final short lease in September 1978. The peeling paintwork on the tail fin is beginning to show traces of its Turkish Airlines scheme, following years of neglect at Marana. Impounded by government authorities in the late 1980s for non-payment of taxes by its registered owners, the 707 began to be robbed of certain valuable parts such as engines, doors and interior fittings, and by October 1993 the aircraft had been moved to a remote site and was sitting forlornly on its tail, with the nose section neatly removed as the scrap men began finally to break the airframe up.

Boeing 707-139 N778PA sits among the assorted scrap in the yard of South West Alloys, Tucson, in October 1992, having been disposed of by the USAF following spares recovery in AMARC as part of the KC-135E re-engining project. This aircraft has led a very full life, having been initially part of an order for Cubana. Though never actually delivered, the order was taken over by Western Airlines where it became N74613 for two years before Pan American bought the aircraft, becoming "Clippor Skylark" N778PA in their expanding jet fleet and converted to a fan-engined B model. The 707 was leased temporarily to THY Turkish Airlines as TC-JBE in 1974, but following sale to lease company Trans Asian in 1976 it would spend the rest of its career moving between numerous different carriers around the world. Between 1976 and 1982 the 707 was flown on behalf of no fewer than eleven airlines, including Bangladesh Biman where it took up the marks S2-AAL. Returned to the lessor in February 1977 it acquired the registration 9G-ACJ and proceeded to be operated by Saudia, Merpati Nusantara of Indonesia, Bahamas World, Aer Lingus, Olympic Airways, Mandala Airlines, Air Malta, British Caledonian and Egyptair before it was sold in August 1980 to another lessor and joined the fleet of Israeli charter airline Maof, whose colours it is wearing in the photograph. During its time with the Israeli carrier the 707 made a starring role in a TV movie true story, re-enacting the Beirut hijacking of an American airliner. The last service it flew for Maof was from Nairobi to Tel Aviv on 7 November 1984, following which the airline ceased trading. Initially stored at Tel Aviv, the 707 flew to Miami in November 1985 after being sold to Aerocar Aviation Corporation, from where it was sold to Boeing Military Airplane Company and flown to Davis-Monthan in March 1986. Sadly, this illustrious airliner was to suffer a pitiful end: the day after it was photographed in October 1992, during dismantling in the contractor's yard, a stray spark from one of the cutter's blowtorches accidentally ignited the interior and the aircraft caught fire and was completely burnt out.

Mohave County Airport, Kingman, Arizona, was famous in the late 1940s as the final resting place of thousands of surplus World War Two combat aircraft, lined up on the acres of desert scrubland awaiting the smelter following the end of hostilities. The quiet airfield, situated at the northern end of the town, has few residents, but has recently begun to fill with many airliners retired by recently-defunct carriers, brought about by the present recession in the airline industry. During the late 1970s and early 1980s British Airways traded most of their surviving 707s back to Boeing in a deal involving the purchase of new 737 and 757 aircraft. The 707s, which were by then almost twenty years old, were of no use to the manufacturer, who subsequently arranged for them to be flown to Kingman for storage, pending their fate. In total, eight former British Airways aircraft were ferried to Kingman between January 1976 and May 1981, and over the next few years they were slowly broken up at this sleepy airfield, alongside a number of very early production DC-8s which had been the property of United Airlines. G-APFO was the fourteenth 707 to be delivered to BOAC out of an initial order of fifteen placed with Boeing in 1956 to augment the Comet 4 fleet on their international route network from Heathrow. The 707 was transferred to the airline's holiday charter division BEA Airtours in 1972, eventually becoming British Airtours two years later when the airlines merged fully. It was finally retired from service in March 1981 and is shown in September of that year parked alongside another ex-British Airways example, both of which had been completely broken up by 1987.

Turk Hava Yollari (THY), the national carrier of Turkey, became a Boeing 707 operator in 1971 when it leased four aircraft to expand its European routes. These very early series 321 models were returned the following year and replaced by another batch of aircraft, this time smaller series 100 models, again on lease. In the late 1970s the airline acquired a number of former Pan American inter-continental 321B/C variants which it flew extensively until their retirement in early 1985, when they were returned to lessor ATASCO Inc. and flown to Stansted Airport, home of well-known 707 contractor Aviation Traders. Here the aircraft received American registrations and spent a few months in storage awaiting an uncertain future; at one time they were rumoured to be in line for possible conversion into tanker configuration for transfer to a military customer. Eventually Boeing purchased the three aircraft and ferried them to the USA, where they joined the ever-growing collection of airframes in Tucson, Arizona. One of the 707s, the former TC-JBU, was broken up whilst in AMARC during the late 1980s, and its remains, consisting of rear fuselage only, were removed by contractor DMI, in whose yard they are seen during October 1992. Of interest is the rather unusual presentation of the US registration, which was applied to the airframe incorporating a non-standard hyphen as N551-7Z.

The perimeter of the huge AMARC, at Davis-Monthan AFB, Tucson, is surrounded by numerous civilian-owned scrap metal contractors, who make a good living disposing of the many former military airframes released for sale by the DRMO. All combat types received by the contractors must be permanently disabled before removal from the site, which generally involves having the wings and tail section severed to avoid any chance of taking flight again and possible acquisition by potentially hostile adversaries. Conversely, certain types are suitable for conversion and sale to civilian operators, such as Convair C-131s as freighters and Boeing C-97s for fire bomber use. DMI Inc. are one such contractor who in recent years have expanded from processing former AMARC inmates into the world of retired civilian types which proliferate in this part of the USA. Such is the nature of the Tucson area, and because of the vast number of dilapidated airframes which abound, the most unusual of sights often becomes the norm. This view, taken in DMI's yard during October 1992, shows the rather bizarre remains of three Boeing 727s which were all considered to be of more value as parts than intact. During the mid-1980s a number of ex-Eastern Airlines Boeing 727s, which had been stored at nearby Marana for a number of years, were broken up, with substantial amounts of their remains being moved to DMI, including some sections of fuselage which have since been ingeniously turned into large storage "sheds". Their nose sections have survived in the yard in this form since then, presumably with some future for them in mind by their owner. Watching over these remains is another 727, formerly flown by Chilean carrier LADECO, which was allocated marks N4367J upon sale to Aeronautical Support International Sales in May 1992, though the aircraft still wears the registration CC-CFG. It became the first whole, ex-civilian airliner to be received directly into one of the contractor's yards without having come via AMARC when it was flown into Davis-Monthan AFB and towed the short distance to DMI on the western edge of the base in the summer of 1992. By October that year its wings had been cut off, and all other valuable components such as engines, interior and undercarriage had also been removed.

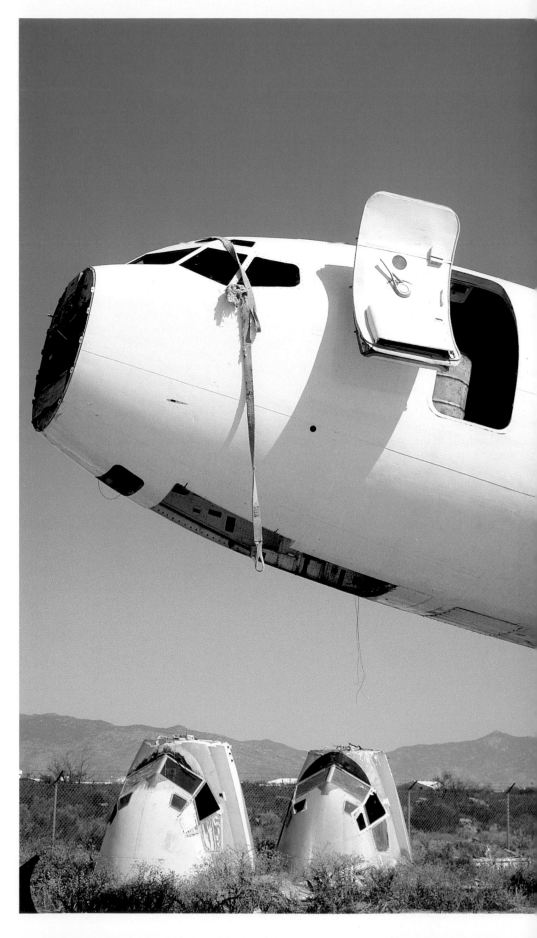

With its once proud owner's name crudely overpainted and many parts either missing or removed, this Boeing 720-023B sits sadly on its wooden trestles amongst scores of similar types, in the Drop area of the vast AMARC. At one time this aircraft was the plush team mount of the famous Los Angeles Dodgers baseball team, wearing the distinctive registration N1R. As such, it was used to transport the team around the USA bearing the name "Kay '0 II" during the seasons between 1971 and 1983. Originally delivered to American Airlines in December 1960, the Boeing ended its flying days in the same way as almost 200 similar examples, donating engines and airframe parts to the US military's KC-135E tanker programme. The Dodgers relinquished their extravagant transport in April 1983, when it was purchased by a company called Great American Airways and sold immediately to Boeing, who flew the aircraft to Tucson for its final duty.

Fifteen years of extreme daytime Arizona temperatures have taken their toll on this Vickers Viscount 798D N776M, whose fading and peeling paintwork gives the aircraft a very unkempt look. One of the large number of Viscounts once owned by the Go Transportation Group, it was photographed at Tucson in October 1992, the one-time base of the company before operations ceased. Since its acquisition by the group in June 1979 this particular Viscount has never been flown, instead spending the entire time parked on the company's maintenance ramp on the north side of Tucson's international airport. Originally laid down on the Vicker's production line as a series 745D for Capital Airlines, the aircraft was never delivered and instead was finished as a 798D model and sold to Northeastern Airlines in October 1958, as N6959C. Following commercial service the aircraft became a corporate transport for a US oil company, and at one time was actually owned by the Confederate Air Force, albeit briefly. The Viscount was acquired by the Go Group in 1979, via a finance company who are the registered owners, and has remained grounded since its arrival at Tucson.

The untidy remains of this Boeing 720-030B were photographed in the yard of South West Alloys, Tucson, in October 1992, after being disposed of from the stocks of stored airframes in the nearby AMARC, where they had donated many valuable components for Air National Guard KC-135E conversions. This 720 was built for Lufthansa in 1961, who operated it as D-ABOL "Stuttgart" in their early days of jet operations, where it was used on the airline's intermediate range services until replaced by the more efficient 727s. Sold to Pan American in 1964, it later moved on to Columbian carrier Avianca who registered it HK-676 in April 1973. It continued to fly for the airline, helping to open up new routes within the Americas as well as trans-Atlantic services to points in Europe, until its sale to Aeron Aviation Corporation in 1983. Registered N3831X, it moved soon afterwards to another broker who in turn sold it to Boeing for use in the KC-135E re-engining project. Gutted, the aircraft is now no more than a basic "hulk", and will be broken up with the help of a huge guillotine blade which will be dropped onto the airframe from one of the contractor's cranes in the background.

As more modern and economical types become available to the world's airlines, so their older and less productive aircraft find themselves squeezed out of the more prestigious routes and onto second-line duties before they are finally retired and disposed of. United Airlines, formed in 1934, is one of the world's largest carriers and recently began to dispose of its early Boeing 737s which have served the airline well since their introduction in early 1968. This example, N9014U, is a very early production airframe, being the thirty-first of the type to be produced. A series 222 model, which boasted a slightly longer fuselage than the original 100s, United were the first operator to take delivery of the larger capacity variant which was produced to cater for the demand by carriers for increased capacity on their shorter "local service" routes. N9014U spent its whole career owned by and flying for United, interrupted only by a short spell on lease to domestic carrier Air California in 1980. By 1992, however, it had been retired and flown to Tucson International Airport, with two other examples which were all undergoing extensive parts reclamation in Hamilton Aviation's compound during October of that year.

Some of the longer term residents at Marana's airliner storage area have begun to be disposed of recently, in a move to tidy up the airfield and create more room for further incoming airliners due to be stored in the coming months. Moved to a more remote part of the airfield, the aircraft dismantlers have started to break up at least two airframes by October 1993 including this former Mackey International Airlines DC-8 series 51 N804E. Originally delivered to Delta Airlines as a series 11 model over twenty-four years ago, Douglas's twenty-fourth production DC-8 acquired Mackey's colourful scheme in September 1979, when it began a twelve-month period of lease with the airline. This sad sight spells the end of the line for this vintage DC-8, which has been at Marana since January 1983 after being repossessed by the owners from lease agent F. B. Ayer & Associates Inc. in February the previous year. As most valuable parts are now removed, the airframe "hulk" will now most likely be broken up for scrap.

Unfortunately, there is no escape from the overcapacity in the airline industry at the moment, as the worldwide recession tightens its grip on air travel forcing operators to trim down their fleets by retiring the older or less economical aircraft in favour of newer more modern alternatives. Some of the relatively modern widebody types to have been withdrawn in recent years are, sadly, not immune from permanent retirement. At Mojave during October 1993 were three Lockheed Tri-Stars which had recently been retired by Japan's largest passenger carrier All Nippon Airways. One aircraft was still parked intact on the airport's main ramp, whilst its two former fleetmates had been towed across the field having been severely cannibalised for spares, as illustrated here.

With all markings and identity painted over, the eighteen-year-old TriStar has suffered from the recovery of valuable airframe parts; with numerous holes cut into the fuselage, the now very dilapidated airframe will surely not last much longer before being finally broken up completely.

As the collected Boeing 707 airframes in AMARC are progressively robbed of all useful components needed for the air force's tanker re-engining programme, together with many other useful airframe parts which have commonality with the C-18/C-137 and E-3 aircraft in the inventory, so the airframes are made available to civilian scrap contractors for final disposal. Periodic sales are held within the AMARC where, through a series of sealed bid auctions, tenders are submitted by the various scrap metal processing companies which surround the vast storage site in Tucson, Arizona, for the available airframes. South West Alloys have been particularly busy in recent years in the acquisition and disposal of many of these former civilian airliners. This October 1992 view of one corner of the busy yard shows three ex-TWA 707s purchased in the air force's sale number 41-2160, where lot #1 was N798TW which arrived at Davis-Monthan on 13 September 1982 and, alongside, N748TW, lot #5 which had arrived in July of that year. Previous scrapyard inmates can be checked by an inspection of the assorted tail fins from the aircraft of earlier sales, which still litter the compound as the photograph shows.

The second-hand market for ageing, noisy "stage 2" airliners has all but disappeared, save for a few regions of the world that can still wring out a justifiable living from original, first generation jet transports. Consequently, the collection of former TWA Convair 880s which are stored at Mojave would seem to have their fate virtually assured: with no examples of the type now flying, prospective customers are nil and suitable spares would seem to be difficult to source. Gulfstream American Corporation, formerly known as American Jet Industries, almost cornered the market on this high performance airliner in the late 1970s, when they purchased the majority of the TWA Convair 880 fleet, which had been withdrawn from service and stored since the mid-1970s, and had them flown to Mojave Airport in 1979. The company had ambitious plans to convert the aircraft into freighter configuration through the installation of a fuselage side cargo door, ahead of the wing on the port side. Coupled with engine "hush kits" to give the type more appeal, they planned to market the aircraft to cargo operators as an alternative to types then available, such as the 707 and DC-8. Sadly, the project was abandoned, and although three aircraft did receive the side cargo door conversion at Mojave, no sales were ever forthcoming and the whole programme was shelved. N803AJ, still wearing its TWA livery and fleet number 8806, is seen at Mojave in October 1990 in the company of more than fifteen similar types.

This aerial shot, taken during October 1993, gives a good view of part of South West Alloys' salvage compound, alongside Davis-Monthan AFB, Tucson, Arizona. The remains of ten Boeing 707 and 720 airframes are visible, all former inmates of nearby AMARC released by the military after all useful parts were removed. Airliners previously operated by Somali Airlines, TWA, Transbrasil, Avianca, Western Airlines and Air Haiti are visible, together with the rear fuselage from an ex-American Airlines example, and at bottom right the wings and tail planes from the Maof 707 which was burnt out in the yard during 1992. The gathered fragments from earlier scrappings litter the yard with the assorted tail fins laid out neatly to one side, giving a clue as to the identity of those aircraft which have been processed through the yard in the past few years. The aircraft, purchased at auction from the base, are simply towed across the main road from AMARC through the large gates on the perimeter fence, top left in the picture.

98

Hamilton Aviation, based at Tucson International Airport, has been carrying out phase inspections and major overhauls for numerous Convair operators for over fifteen years. Their in-depth knowledge of the entire family of twin-engined Convairs is backed up by an immense spare parts inventory, available for in-house overhaul programmes as well as customer requirements. In order to attain this level of service and support, the company purchased a number of Convairs for parts, most of which were broken up at Tucson over the years. One example which has survived is this 580 model N5822, which still wears the full markings of its last operator Atlantic Gulf, a short-lived commuter carrier based at Clearwater Airport, Florida. Photographed at Tucson in October 1993, this aircraft has been present in Hamilton's compound since the late 1980s, and although relatively intact it has donated many useful parts including engines and interior fittings.

The one that didn't quite make it! Boeing 707-321B N320MJ crashed on take-off at Marana, killing the co-pilot and seriously injuring the other two crew members on a short flight to Davis-Monthan AFB on 20 September 1990. Miraculously, the wreckage managed to miss the many stored airliners which were parked nearby on the field as it came down in an open area close to the runway. A one-time Pan American N891PA, the 707 was later flown by British West Indies Airways as 9Y-TEZ until its sale in 1982 when it was registered as N3127K to FTC Watson Wings Inc. Further ownership changes followed with the 707 becoming VR-CBN, before ending up at Santa Barbara Airport, where it was stored in April 1985. The aircraft's condition deteriorated over the next five years as it sat in the corrosive atmosphere of this oceanside airport in California where it earned the nickname "The Shadow", unofficially donating many parts to other 707s which were on maintenance or under conversion on the airfield at the time. Registered in September 1990 to Omega Air, a broker for many of the 707s bought back by Boeing on behalf of the USAF, it was restored to flying condition and flew to Marana soon afterwards. On the fateful day, the 707 was being flown the short distance to Davis-Monthan on what was to be its last flight anyway, to join the other similar types in AMARC. As the aircraft became airborne on the southern runway, its starboard wing dropped and struck the ground; the aircraft cartwheeled, crushing the flight-deck, before the fuselage broke forward of the wing. The accident report later stated that the extreme lack of basic flight-deck instruments in the aircraft at the time caused crew disorientation once it had rotated, as the horizon disappeared below the nose, giving no clue to the aircraft's attitude.

From humble beginnings in the mid-1960s as a small helicopter operator based in the north-western United States, Evergreen has developed into one of America's major supplemental freight and charter carriers, whose yearly revenue now runs into hundreds of millions of dollars. The airline operates a very varied assortment of types, flying a mixed fleet of Boeing 727, 747, Douglas DC-8 and DC-9 aircraft on passenger and cargo services, as well as smaller types such as Casa 212s and numerous helicopters which are used in support of various government support programmes around the world. Over the years the company has branched out into other markets, including the operation of a fleet of air tanker aircraft during the 1970s and early 1980s on contract to the United States Forest Service (USFS). Five former Navy Lockheed SP-2E Neptunes were acquired from military stocks at Davis-Monthan during March 1977 to operate alongside a converted B-17G Flying Fortress. Three of the Neptunes were converted into fire bombers and were given the company's attractive green and white livery, with the other two airframes being held in reserve as a source of spares. The aircraft continued to serve in this role until the mid-1980s, when they were recalled to their Marana base and withdrawn from service after Evergreen came out of the fire bomber business. Tanker "145", N206EV, is shown in retirement at Marana in October 1991 flanked by other Neptunes sitting quietly awaiting a decision on their future. One of the tanker aircraft has recently been pulled from storage and had its Turbo-Compound engine "de-pickled" in preparation for its forthcoming restoration to original US Navy configuration by the company's vintage aircraft division, Evergreen Ventures.

This Fairchild F-27 N5098A was photographed in October 1992 alongside Interstate 10 at Quartzsite, a small town in western Arizona midway between Phoenix and Los Angeles. Quite literally a "desert airliner", the F-27 had been bought by a local resident who had it trucked to his Quartzsite house from Pinal Air Park, Marana. Here it joined a collection of similarly derelict airframes including a Beech 18, numerous light twins and a Cessna 172 perched on top of the house roof. The aircraft had last flown for Columbian carrier TAC, whose scheme it still wears, who had leased it along with two other examples from F. B. Ayer & Associates in the early 1970s. Returned to its owners, the F-27 was flown to Marana and placed in storage in the early 1980s. Over the next decade the aircraft's condition grew steadily worse as it was robbed of many parts, and by 1990 it had eventually been dismantled. The aircraft consisted of no more than a bare fuselage and wings when purchased in early 1992, and is shown not long after arriving at Quartzsite still on its trailer.

One of the more unusual residents at Tucson International Airport for the last sixteen years has been this Caravelle VIR N777VV. The world's first short to medium range turbojet airliner never enjoyed much success in North America, with only United Airlines operating it in significant numbers. The type was also flown by Midwest Air Charter, who later sold their six examples to Airborne Express, the Wilmington, Ohio-based freight carrier. Originally part of United's fleet, operated as N1002U, it was sold in 1970 along with six others to Transavia Holland, where it took up the registration PH-TRY. It flew with the Dutch carrier for the next six years before it crossed the Atlantic once again when it became the property of Independent Air. In an unusual move, it was acquired by Go Transportation in 1977 and delivered to Tucson, where it has remained grounded ever since. Stripped of any markings the Caravelle now sits forlornly in the desert, unlikely ever to fly again.

BETTER TIMES AHEAD

A sad manifestation of the recession in the airline industry is the numbers of brand new airframes which have joined the desert airliners in the last few years. Known in the industry as "white tails", these airliners have yet to take up any carrier's colours, having come directly from the manufacturers' production lines. The healthy markets of the 1980s, when airlines were ordering much new equipment, gave way to a massive downturn in passenger traffic during the early 1990s, the result of worldwide recession and overcapacity. This reduction in demand by the airlines coincided exactly with the manufacturers' peak production rates, a consequence of the earlier order levels, which led to the extraordinary situation of factory-fresh aircraft being put directly into storage. Mojave during October 1991 had this line-up of five brand-new McDonnell Douglas MD-83s, which had arrived from Long Beach the previous month. Owned by leasing company Irish Aerospace Ltd, the airframes found work soon afterwards with the Columbian carrier Avianca, and all had departed on lease by April 1992.

Not all of the desert 747s are trade-ins or retirements. This immaculate series 400 model was flown to Mojave in late 1991 for storage, never having carried a fare-paying passenger. C-GAGL was one of three brand-new 747-433 Combis which were built for Air Canada and handed over in June 1991. Not immediately needed by the airline, they were initially stored at the manufacturer's Everett plant before being ferried to California a few months later. Suitably protected against the heat and dust which prevail in this part of the world, the 747 has had it windows and doors sealed, and the engine intakes covered over to preserve the aircraft in mint condition. Better times were ahead for the aircraft, when on 20 March 1993 it was flown to Las Vegas and on to Toronto in preparation for its introduction into service. It finally entered service with Air Canada on 29 April on the carrier's Toronto to London-Heathrow schedule.

The one that started it all off. Boeing's model 367-80, a radical design at the time, proved to be the pattern for so many other passenger aircraft which were to follow in its wake, and gave Boeing the undisputed lead in commercial aircraft design and manufacturing for years to come. The aircraft was originally designed to offer the USAF a suitable jet-powered conversion of their KC-97 tanker aircraft fleet, which was beginning to show its unsuitability in contrast with the air force's recently introduced jet-powered B-47s. In order to maintain the secrecy of the project Boeing continued to refer to it by the KC-97 model number, the Boeing 367, and it was known as the 367-80. The knowledge gained by Boeing in the B-47 project was to pay handsome dividends in the design of the new jet transport, which would later emerge as a whole new aircraft. The programme was self-funded by Boeing to the tune of $16 million, quite a gamble for a private company at the time, though considerable interest in the aircraft by the government and the air force helped to convince the manufacturer they were doing the correct thing. The design evolved into the successful KC-135 tanker, from which the world-beating 707 was later developed, even though the aircraft differs quite significantly in a number of areas from the tanker. Since its first flight in July 1954, "Dash 80" N70700 has been used to develop all subsequent variants of the 707 and 720, during which time it acquired many non-standard pieces of equipment and carried out many harsh flight trials. Honourably retired by Boeing after a life of 1,691 flights over a total time of 2,350 hours, the aircraft was flown to Davis-Monthan AFB in August 1976 to be held in storage on behalf of the Smithsonian Institute, who recognised the "Dash 80" as being one of the twelve most significant aircraft of all time. Still wearing its original cream-yellow and brown colour scheme, and slightly incorrectly titled a 707, the "Dash 80" was photographed at Davis-Monthan in 1982, parked on what is known in AMARC as "Celebrity Row", where one example of every type presently held in store on the base is displayed. By 1991 the historic aircraft had been "de-preserved" and was flown out of AMARC to begin its well-deserved retirement on display at the Smithsonian Museum.

The ranks of stored airliners in the deserts of the south-western United States hold one or two significant airframes which are richly deserving of a more secure future. As mentioned earlier, Boeing's commercial airliner progenitor, the Dash 80, spent time in storage at Davis-Monthan AFB before joining a national museum collection. Ironically, less than fifty miles away from the Dash 80's one-time storage site, Douglas's equivalent also sits in open storage awaiting an uncertain future. Douglas DC-8 XA-DOE has been parked at the Pinal Air Park, Marana, for the last fourteen years following return from lease

Probably the second most significant airliner to be stored in the desert, after Boeing's Dash 80, came from the same stable. The first Boeing 747, N7470, was photographed in storage at Las Vegas-McCarran Airport during September 1981, not long after being withdrawn from use by the manufacturer. The aircraft first flew on 9 February 1969 and was built to production model standard, there being no prototype 747 ever produced. N7470 spent

Another manufacturer's "first" which is seemingly abandoned after its retirement from commercial service. Douglas DC-9 line number #1, N914LF, was photographed at the huge former military airfield of Sherman-Grayson in northern Texas during October 1993. As the picture shows, most useful airframe parts have been removed, including both engines, the left main gear and one of its starboard wheels. This historic aircraft sits amongst a number of similar types, all of which have undergone extensive parts recovery with some beginning to be broken up for scrap. First flown on 25 February 1965, wearing the appropriate registration N9DC, it was delivered to Texas International the following year and re-registered N1301T. In 1983 it was sold and began a life of short leases to numerous different operators both in the United States and Europe. Its last passenger-carrying services were in summer 1991 on behalf of Viscount Air Services, by which time it had acquired the registration N914LF, following which it was bought back by its manufacturer the McDonnell Douglas Corporation. This move could have secured a safe future for the airliner, but in May 1992 aircraft spares supplier Aircraft Support Group Inc. purchased the airframe and flew it to Clearwater, Florida, for storage. Its final assignment came soon after, when the DC-9 played a dramatic role in a feature film, which involved lowering a stuntman on a wire from its tail cone, who then proceeded to climb into a Jetstar executive jet whilst airborne. Finally retired to Sherman-Grayson, one can only hope someone recognises this DC-9's heritage.

As early as 1981, some airliners were being produced faster than their customers could accept them. This DC-10-30, photographed at Marana in September of that year, was flown directly into storage from Long Beach, still in its bare metal finish. Originally ordered by National Airlines as N84NA, the aircraft had not been completed when the airline was merged with Pan American during 1980, and consequently, on its completion in August of that year, it was stored temporarily pending delivery to its new owner. The DC-10 was eventually delivered to Pan American on 17 October, when it departed Marana for Miami where it was painted into full Pan Am scheme, and operated for the next three years as "Clipper Glory of the Skies". Bought by United Airlines in April 1985, it was re-registered into their fleet sequence as N1855U, which it still wears today.

As well as receiving Boeing 707 and 720 airframes to provide parts for the USAF KC-135E programme, AMARC contains a number of 707 aircraft from civilian sources held temporarily for Grumman's E-8C J-STARS programme. J-STARS (Joint Surveillance and Target Acquisition Radar System) is a co-funded US Army/US Air Force project which offers a battlefield surveillance platform that can locate, identify, classify and track hostile ground targets whilst operating in a "stand-off" position. Grumman's Melbourne System Division was awarded the contract to adapt suitable airframes for the task, which fell to the ubiquitous Boeing 707. In a money-saving move, and due to the closure of the 707 production line, a number of former airline examples were acquired in the late 1980s on behalf of the programme. Flown to AMARC, the aircraft are held in storage awaiting their turn on the Melbourne conversion line. Of the earlier 707 aircraft stored at Davis-Monthan, the last twenty-four airframes to arrive are held as parts donors specifically for the J-STARS programme. This immaculate-looking example, a 1966 vintage 384 model, last flew for Royal Jordanian as JY-AEC. Originally built for Olympic Airways as SX-DBB, the 707 was bought from Royal Jordanian by Omega Air in June 1992 and was ferried from Amman to San Antonio, Texas, via Shannon on 15 September that year. It is shown at Davis-Monthan in temporary storage the following month.

Even with titles, logo and registration overpainted there is no disguising this Boeing 767 as belonging to British Airways, one of a number of the type which spent a short time in store at Mojave during 1991. An extended range series 336ER, G-BNWJ "City of Athens" was rolled out at Paine Field, Everett, on 12 March 1991, its first flight coming sixteen days later. Handed over to British Airways on 24 April, the 767 was flown to Mojave where it was temporarily stored while the airline made preparations for its entry into service. The type was to replace ageing TriStars on the carrier's New York routes from Manchester and Glasgow, as well as operating on high-capacity short range services such as Heathrow to Paris. The 767 order from British Airways involved the trading back to Boeing of a number of TriStar airframes, which the type would eventually supercede. These aircraft were ferried to Mojave for storage and eventual disposal by Boeing, which will most likely entail the aircraft being broken up and sold for parts. As can be seen in the photograph, taken on 4 October 1991, the 767 was being prepared for flight at Mojave. It eventually departed its temporary home in the desert on 19 October when it was ferried to New York-JFK and entered service soon afterwards on a London-bound service.

Some of the newest aircraft presently stored have already led an interesting short life, such as this Boring 767-231 N607TW which was delivered new to TWA on 26 July 1983. In a move to avoid bankruptcy, some financial restructuring by the airline resulted in the aircraft being sold in August 1990 to First Security Bank of Utah, though still operated by TWA. However, on 8 September the following year the 767 was withdrawn from service and flown to Kansas City for storage with a total airframe time of 30,800 hours in 8,602 flight cycles. Here it remained until early October, when it was registered to United Aviation Services and flown to Marana, Arizona, for further storage, where it is seen on 9 October in a basic TWA style scheme. Transferred back to First Security Bank of Utah, it was ferried to Los Angeles on 6 November and leased to TWA once again, entering service with the carrier four days later.

Even though the notice warns that Mojave Airport's north side is out of bounds, its many occupants are all too obvious to the casual observer. At the time of this October 1993 photograph, there were no fewer than eighteen widebody airliners in storage on the airfield, including nine TriStars, one DC-10 and A-300, seven different Boeing 747s and two brand-new MD-11s, together with almost one hundred other assorted smaller types. The MD-11 which sits at the head of this line is awaiting service entry with Delta Airlines, having been delivered into storage directly from the production line at Long Beach. As the airframe is not actually owned by the airline, it is the finance company who will suffer through its extended time in storage. This is now the norm: newly-built aircraft being the property of financiers and lease companies who have arranged lease packages to cover most major airlines' re-equipment plans.

The shimmering heat haze across the Mojave runway gives the impression of a mirage, though this view of three brand-new series 400 747s stored in the desert is most definitely tangible. The three series 433 Combi aircraft, C-GAGL, C-GAGM and C-GAGN were manufactured in 1991 for Air Canada, having been ordered during the more buoyant 1980s, and following roll-out at the factory were all immediately placed into storage at Everett, temporarily unwanted by the airline. The 747s were later ferried to Mojave where they were prepared for an extended period of inactivity, and were photographed in store during October that year. The eventual need for increased capacity by the airline on its trans-Atlantic routes led to the aircraft being put into service in early 1992, with C-GAGL performing its first revenue-earning flight on 29 April between Toronto and London-Heathrow.

A desert airliner, though not commercially owned, is this Boeing 727-30 which operated for the US Air Force as 84-0193 under the military type designation C-22A. One of six C-22s operated by the USAF, all former civilian airliners, this aircraft is the sole "A" model variant which was acquired for use as transport for the Commander-in-Chief of US Southern Command, and was operated from Howard AFB, Panama, by the 310th Airlift Squadron. The C-22A was originally delivered to Lufthansa as D-ABID in April 1964; later bought back by Boeing it was eventually transferred to the Federal Aviation Administration (FAA) where it took up the registration N78 in 1976. The USAF added the type to its inventory in 1983 when N78 took up its military identity, followed in 1985 by four more former Pan American airframes which became C-22Bs. A single former Singapore Airlines series 200 B.727 was later added, which was given the "sub-type" C-22C. 84-0193 was finally withdrawn from use by the USAF in November 1991 and flown to Davis-Monthan AFB for storage in the AMARC. Photographed here the following October, the aircraft is parked on "Celebrity Row" wearing the simple military scheme adopted by the type and United States of America titles.